Books by James DeMeo

* *The Orgone Accumulator Handbook: Construction Plans, Experimental Use, and Protection Against Toxic Energy,* Natural Energy Works, Ashland, Oregon, 1989.

* *Saharasia: The 4000 BCE Origins of Child Abuse, Sex-Repression, Warfare and Social Violence In the Deserts of the Old World,* Revised Second Edition, Orgone Biophysical Research Lab, Ashland, Oregon, 2006.

* (Editor) *Heretic's Notebook: Emotions, Protocells, Ether-Drift and Cosmic Life-Energy, with New Research Supporting Wilhelm Reich,* Orgone Biophysical Research Lab, Ashland, Oregon, 2002.

* (Co-Editor with Bernd Senf) *Nach Reich: Neue Forschungen zur Orgonomie: Sexualökonomie, Die Entdeckung der Orgonenergie,* Zweitausendeins Verlag, Frankfurt, 1997.

* (Editor) *On Wilhelm Reich and Orgonomy,* Orgone Biophysical Research Lab, Ashland, Oregon, 1993.

* *Bibliography on Orgone Biophysics: 1934 to 1986*
Updated on-line version 1920 to Present now available at:
www.orgonelab.org/bibliog.htm

* For a full listing of Dr. DeMeo's publications:
www.orgonelab.org/demeopubs.htm

PRELIMINARY ANALYSIS OF CHANGES IN KANSAS WEATHER COINCIDENTAL TO EXPERIMENTAL OPERATIONS WITH A REICH CLOUDBUSTER

With the Appendix,

Evidence for a Principle of Atmospheric Continuity

And a New Preface, 2010

A Research Thesis
Geography-Meteorology Department
University of Kansas at Lawrence, 1979

by
James DeMeo, Ph.D.

Director
Orgone Biophysical Research Laboratory
Ashland, Oregon, USA
www.orgonelab.org www.saharasia.org

First Published Edition 2010

Published by

Orgone Biophysical
Research Laboratory, Inc.
Ashland, Oregon, USA

Worldwide Distribution by:
Natural Energy Works
PO Box 1148
Ashland, Oregon 97520
United States of America
www.naturalenergyworks.net
info@naturalenergyworks.net

and

Lightning Source

Preliminary Analysis of Changes In Kansas Weather Coincidental
to Experimental Operations With a Reich Cloudbuster

For information, send inquires to the above address.

ISBN: 978-0962185540

10-digit ISBN: 096218554X

First Published Edition, with New Preface, 2010

1 2 3 4 5 6 7 8 9

TABLE OF CONTENTS

LIST OF TABLES AND FIGURES

Figure Page

Tables

In Appendix I:

LIST OF PHOTO SEQUENCES

Preface to the 2010 Edition

This work is being republished from a little-known 1979 academic research document, for both its scientific and historical merit. It summarizes my early experimental investigations into the atmospheric discoveries of the late Dr. Wilhelm Reich, which constituted, then as now, a major scientific discovery and controversy. It also was the first research to be openly undertaken on any aspect of Reich's biophysical findings in any university, following years of unethical smear-attacks against his work and person, in magazines and medical journals only some 30 years earlier. Those assaults led to his death in prison, and an unprecedented US Court order to literally *ban and burn* his scientific publications. As such, and given the positive results I obtained which confirmed some of Reich's most unorthodox claims, my work constituted as much of a social breakthrough as a scientific one. This was in no small part due to the unusual personal support I enjoyed at the University of Kansas, Geography-Meteorology Department, where this work was undertaken.

This study focused upon the *cloudbuster* device, a type of antenna employing the techniques of what Reich more formally called *Cosmic Orgone Engineering*, and one of his very last and most controversial discoveries. Comparatively speaking, the cloudbusting method is something like *atmospheric acupuncture*, where a special antenna "needle" is used according to specific principles to alter the movement of energetic currents and charge within the atmosphere, following which one may observe changes in circulation, humidity and cloud dynamics. The cloudbusting methods and their atmospheric effects are so flagrantly inexplicable from the viewpoint of contemporary and orthodox scientific concepts, that any positive evidence for its functioning carries the most profound implications. Positive evidence on the cloudbuster would give added strong support to Reich's whole line of research which led to its development, such as with the *orgone accumulator* and the *orgone energy* itself. And this is heresy. It suggests that cosmic space and our atmosphere is filled with an excitable and pulsatory energetic continuum (see Appendix I), which affects or even controls our weather.

One of the problems I faced in the late 1970s was finding a university where I could openly study and research this highly unorthodox subject. It required about two years of searching, but I finally connected with several professors at the University of Kansas (KU) who were sympathetic and open to my ideas. I enrolled in the KU Geography-Meteorology Department's graduate program, to study classical earth and atmospheric sciences (meteorology and climatology), mixed with other courses on oceanography, geology, social science, cartography and astronomy. Geography was interdisciplinary, and well-suited for any study of Reich's own interdisciplinary findings. With agreement from my professors, I undertook new experimental work investigating and attempting to replicate Reich's original claims, without bias or prejudice, pro or contra.

At KU, I was especially helped by Prof. Robert Nunley, who acted as Chairman for my research program, and also became a good personal friend. He helped to secure my position as Teaching Assistant and later Instructor in the Department, where I eventually proposed and taught my own courses. I developed a new curricula for courses on *Renewable Energy Systems*, and a more controversial *Rhythms and Cycles in the Natural World*, where the findings of some of the scientists mentioned in Appendix I of this publication were presented, including those of Reich.

Prof. Nunley also invited me to develop an atmospheric research station at his farm just outside of town, where my large cloudbuster antenna was set up along with weather monitoring apparatus. That is where much of the research described in this book took place. I attended classes at KU and looked after my teaching responsibilities, doing research in available free time. An experimental protocol on the cloudbuster was worked out in consultation with the professors on my committee, and others in the Geography-Meteorology Department.

During this same period, I also remained in consultation with experts in orgone biophysics, my mentors in orgonomy such as Dr. Richard Blasband in Pennsylvania, and Mr. Robert Morris in Florida. My work would follow Reich's original protocols and safeguards, even while being tested out via the most rigorous of natural scientific methodology. A program was developed for testing of cloud-dissipation effects, as well as for increasing rains, and another for decreasing them. Precautions and safeguards against destructive weather reactions were instituted, at every step, in coordination with the weather forecasting lab at KU.

As I previously saw when assisting Dr. Blasband at his own laboratory in Pennsylvania, and later during my own Kansas operations, some of the weather reactions to the cloudbuster were quite spectacular. However, as my professors cautioned, such weather was typical for Kansas. They would half-way joke with me: *"Jim, you can point your finger into the sky, and sooner or later something is going to happen"*, they would say. *"Yes, but if you do this ten times, and each time it rains within an hour, then we might hire you out as an authentic rainmaker"*, I would respond, usually to some laughter. But it was serious business, and the research protocol was set up with controls and data analysis that would sort out what was truth versus illusion. Some other members of the faculty thought it was a big waste of time, that all I would do was to "prove a negative result", that the cloudbuster did not work. But I proceeded with the experiments in good faith, that the truth would win out, one way or the other. Personally speaking, I had no interest to waste my time chasing illusions. If the cloudbuster could not show good results when subjected to a more systematic analysis than had previously been undertaken, then I wanted to know about it. If it did, however, then Reich was correct, and it really was a major and authentic scientific breakthrough.

At the end of two summers of field work, in 1977 and 1978, it was absolutely clear that the cloudbuster operations were — by even a modest appreciation —

stimulating reactions within clouds and rainfall patterns at the time of the testing. My findings were organized, written up and reviewed by my committee of professors. Weather data from the entire network of Kansas rain-gauges was used to evaluate for results, along with data I acquired from the field site. The present publication contains the original and full documentation, which was formally accepted by the Department in 1979 as a Master's Thesis. The results were excellent and very much in keeping with Reich's claims that the cloudbuster actually worked, firstly to modify individual cumulus clouds, and secondly to influence weather across the entire state of Kansas. Also a special Appendix, *Evidence for the Existence of a Principle of Atmospheric Continuity*, detailed other research by many different scientist suggesting the existence of something quite similar to the orgone energy of Reich. That is also included in this publication.

For their courage and openness to new ideas — which should be a hallmark of true university life and study — I thank those honest professors and loyal friends in the KU Geography-Meteorology Department who supported my rights to do this work, and who in spite of the oftentimes considerable controversy and difficulties it created for them, did not avert their eyes when the evidence supported the unorthodox and "unthinkable". And this includes some of the students. I am informed, the KU Geography Student Bulletin Board is today graced with a photo of my cloudbuster Icarus, with the caption *"May the Force Be With You"*. I take this all in good humor, though it is actually quite closer to the truth than they may wish to consider. George Lucas apparently has also read Wilhelm Reich.

Eventually I completed a PhD at Kansas with new research focused not on the cloudbuster, but on an equally controversial subject related to Wilhelm Reich's findings, as summarized in my book *Saharasia: The 4000 BCE Origins of Child-Abuse, Sex-Repression, Warfare and Social Violence, In the Deserts of the Old World.* Ultimately I went on to serve as professor in four different universities over around 10 years — KU, Illinois State University, University of Miami, and University of Northern Iowa — but always with controversy and personal difficulties due to growing malicious attacks and dirty-tricks by the organized "skeptics". I continued doing open research with the cloudbuster, notably practical experiments for drought-relief in the USA. The positive results from that work provided additional support for Reich's original claims, as well as great social benefits in the ending of droughts. This rallied much support to my side, while my critics were sent off to plot and gnash their teeth. I nevertheless tired of being assaulted for every new and positive step which was taken. The controversy also made acquisition of any permanent university post impossible. This prompted the development of my own private nonprofit research institute, the *Orgone Biophysical Research Laboratory*, which today is located in rural Ashland, Oregon, and through which my research and educational seminars are organized. While the malicious attacks did not stop, without an academic job to protect, I could mostly ignore them. That interesting history is for another article, on another day.

Throughout this period and in the years after my last university posting in 1988, I privately organized and directed various cloudbusting field projects in the USA, to end severe droughts, and with additional investigations into the question of desert-greening. This included extensive drought-abatement work in the SE, Midwest and Western USA, as well as overseas in Greece, Germany, Israel, Namibia and Eritrea. These projects were initiated mostly through invitations from farmers and other concerned individuals living and suffering under drought conditions in their respective regions, and typically funded by private donations. My overseas projects also frequently proceeded by invitation from the agricultural and hydrological divisions of the respective governments. These were always team efforts, and I thank the many different sponsors, associates, and volunteers.

The full history of my cloudbusting field operations, the people and places, the atmospheric and social reactions to them, is the subject of another book in preparation (*The Greening of Droughts and Drylands*). For now, I merely emphasize how the results of more than 30 years of working with this new technique provides exceptional support for Wilhelm Reich's original findings and claims.

This publication provides my own starting point of serious investigation into the subject. More exacting details on my later work are found in the various article publications listed in the section to follow. And I should mention, there are other hard-working and skilled naturalists and scientists who did similar work well before myself, who continue with similar work today, and from whom I learned a great deal. While I never met the late Dr. Wilhelm Reich, much was learned from his original publications (which should not to be confused with the terrible distortions of his work as typically found today on global internet).

From my own years of study and field work on this question, I can state categorically, *the cloudbuster works,* and can be applied for a multitude of drought, heat-wave and wildfire emergency conditions, for the betterment of nature and humankind. It can also work even under the most extreme of drylands or drought-desert conditions. The reader is invited to review the various publications cited in this document, and specifically to review the informative weblinks given below.

To the above I can also point out how certain sub-disciplines of atmospheric science and physics of today have made indirect confirmations of a same or similar energy principle in nature, and speak a language which is closer to that of Wilhelm Reich than they might realize or admit. Astronomy today speaks about an *interstellar* or *intergalactic medium*, or about *cosmic plasmas* or a mysterious *dark matter* — all of these are quite similar in nature to both the 19th Century *cosmic ether of space*, as well as to Reich's mid-20th Century *cosmic orgone energy*. As I detail elsewhere in this publication, as well as in separate papers, the textbooks typically misreport the old ether-drift experiments, some of which did in fact verify its existence, and also suggested it was quite similar in properties to Reich's cosmic orgone.

The field of solar-terrestrial physics today also accepts there are profound weather and climate changes correlated to sunspots, solar flares and other solar variations, though the mechanism has often proven elusive. The field of auroral physics accepts a linkage between solar wind and flares to strong electrical and geomagnetic currents. These give rise not only to the aurora, but also to correlated weather phenomena. Mysterious luminous phenomena such as the *thunderstorm sprites* are hardly less fantastic than any claim by Reich about luminous orgone energy anomalies, described decades earlier. And newer findings on downward-directed stratospheric winds, affecting both jet streams and tropospheric winds, uncannily invite us to look for some kind of streaming cosmic phenomenon which puts the whole of the atmosphere into motion. This is all very much in keeping with Reich's discussions on energy streams which move down to the Earth's surface from space, to govern the formation and movements of large weather systems. His writings on this subject pre-date declassified discussions on the jet streams, just as his ideas on the Earth's blue-glowing *orgone energy envelope* predate the astronaut photographs of the same. Reich's empirical discussions about atmospheric haze, acidity and ozone affecting the health of forests and creating a toxic blackening of open rock surfaces, which his 1950s critics called "coo coo", predated any similar discussion on "smog" pollution by at least a decade. His findings on haze-producing stagnant *drought-haze* or *drought clouds* were 30 years ahead of similar discussions on atmospheric aerosols, "pollution clouds" or the problems of contradictory ocean "dry fog" haze, or desert-haze as acknowledged today. There is much that modern science can learn from Reich, and from my own experiences. Before this is possible, however, they will firstly have to throw out the anti-scientific "skeptic-club" book-burners and ridicule-hyenas from their current high-status positions within the scientific institutions and editorial boards.

Presented here is the original Thesis on the Reich Cloudbuster as it was approved at KU in 1979. Some of the original color photos were degraded and so do not reproduce very well in black-and-white, but they remain discernibly valuable. The text has been freshly typeset and reformatted, and other than correction of a few typos, remains the same as in the original. This new Preface has been added, along with a following cautionary statement about untrained and casual experimentation with the cloudbuster. Because of this issue, I have deleted one section which shows the fine details on how a cloudbuster is constructed, and a few other points on its operation. That information is too critical and dangerous to be spread outside of the current apprentice-mentor system by which it has so far been safeguarded. I've also added a few additional comments to the Appendix document, *"Evidence for the Existence of a Principle of Atmospheric Continuity"*, on pages 108 and 128-129, giving a bit of update. Otherwise, the main body of the Thesis remains unchanged.

James DeMeo, PhD
Greensprings
Ashland, Oregon, USA
June 2010

For More Information:

*Research Summary article on Cloudbusting:
www.orgonelab.org/ResearchSummary2.htm

* James DeMeo's Research Websites:
On the Orgone Energy Question: www.orgonelab.org
On the *Saharasia* Discovery: www.saharasia.org

* *So, You Want to Build a Cloudbuster?*
www.orgonelab.org/sobuildaclb.htm

* *About the Growing "Orgone Nonsense" on Global Internet*
www.orgonelab.org/orgonenonsense.htm

* *A Dynamic and Substantive Cosmological Ether*
www.orgonelab.org/DynamicEther.pdf

* *Bibliography on Orgonomy: 1920 to Present*
www.orgonelab.org/bibliog.htm

Citations for the 2010 Preface Materials

James DeMeo: *Preliminary Analysis of Changes in Kansas Weather Coincidental to Experimental Operations with a Reich Cloudbuster,* James DeMeo, M.A. Thesis, Univ. of Kansas, Geography-Meteorology Dept., 1979.

James DeMeo: "Water Evaporation Inside the Orgone Accumulator," *Journal of Orgonomy*, 14:171-175, 1980.

James DeMeo: "Field Experiments with the Reich Cloudbuster: 1977-1983", *Journal of Orgonomy,* 19(1):57-79, 1985; Greek translation in the *Hellenic Orgonomic Association Journal,* Thessaloniki, 1990.

James DeMeo & Robert Morris: "CORE Progress Report #13, Fighting the Extreme Drought of Spring 1985 (Southeast)", *Journal of Orgonomy*, 19(2):265-66, 1985.

James DeMeo & Robert Morris: "CORE Progress Report #14: Possible Slowing and Warming of an Arctic Air Mass through Cloudbusting", *Journal of Orgonomy,* 20(1):120-125, 1986.

James DeMeo: "So You Want to Build A Cloudbuster?" 1986, Internet Publication: http://www.orgonelab.org/sobuildaclb.htm

James DeMeo: *On the Origins and Diffusion of Patrism: The Saharasian Connection,* James DeMeo, Ph.D. Dissertation, Univ. of Kansas, Geography Dept. 1986.

James DeMeo: "Nine Years of Field Experiments with a Reich Cloudbuster: Positive Evidence for a New Technique to Lessen Atmospheric Stagnation and Bring Rains in Droughty or Arid Atmospheres", *Abstracts of Papers, Program of the 1987 Meeting of the Association for Arid Lands Studies*, El Paso, Texas, p.6.

James DeMeo & Robert Morris: "CORE Progress Report #15: Breaking the 1986 Drought in the Eastern U.S., Phase 3: A Cloudbusting Expedition Into the Southeastern Drought Zone", *Journal of Orgonomy,* 21(1):27-41, 1987.

James DeMeo & Robert Morris: "Preliminary Report on a Cloudbusting Experiment in the Southeastern Drought Zone, August 1986", *Southeastern Drought Symposium Proceedings,* March 4-5, 1987, South Carolina State Climatology Office Publication G-30, Columbia, SC, 1987.

James DeMeo: "Reduction of Rainwater Acidity Following the End of the 1986 Drought: An Effect of Cloudbusting?", *Journal of Orgonomy,* 21(2):249-251, 1987.

James DeMeo: "Field Experiments with the Reich Cloudbuster: Positive Evidence for a New Technique to Lessen Atmospheric Stagnation and Bring Rains in Droughty or Arid Atmospheres", Presented at the Annual Meeting of the *Association for Arid Lands Studies,* April 1987, Western Social Science Association, El Paso, Texas, see Abstracts of Papers, p.6.

James DeMeo: "A Dynamic Biological-Atmospheric-Cosmic Energy Continuum: Some Old and New Evidence", *Geo-Cosmic Relations: The Earth and its Macro-Environment,* Proceedings, First International Congress on Geo-Cosmic Relations, 19-22 April 1989, Amsterdam, Netherlands, G.J.M. Tomassen, et al, Editors, PUDOC Science Publishers, Wageningen, 1989.

James DeMeo: "Response to Martin Gardner's Attack on Reich and Orgone Research in the Skeptical Inquirer", *Journal of Orgonomy,* 23(1):126-133, 1989; also in *Pulse of the Planet* 1(1):11-17, 1989. Internet posted: http://www.orgonelab.org/gardner.htm

James DeMeo: "Postscript to the Food and Drug Administration's Scientific Evidence Against Wilhelm Reich: Results of a Freedom-of-Information-Act Inquiry.", *Pulse of the Planet,* 1(1):18-23, 1989.

James DeMeo: "The Orgone Energy Continuum: Some Old and New Evidence", *Pulse of the Planet,* 1(2):3-8, 1989.

James DeMeo: "Research Progress Report: The Desert-Drought Map", *Pulse of the Planet,* 1(2):82, 1989.

James DeMeo: "Desert Expansion and Drought: Environmental Crisis, Part 1", *Journal of Orgonomy,* 23(1):15-26, 1989.

James DeMeo: "CORE Progress Report #20: Breaking the Drought Barriers in the Southwest and Northwest USA", *Journal of Orgonomy,* 23(1):97-125, 1989.

James DeMeo: *The Orgone Accumulator Handbook: Construction Plans, Experimental Use, and Protection Against Toxic Energy,* 2nd Revised Edition, Natural Energy Works, Ashland, Oregon 1989.

James DeMeo: "The Origins and Diffusion of Patrism in Saharasia, 4000 BCE: Evidence for a Worldwide, Climate-Linked Geographical Pattern in Human Behavior", *Kyoto Review,* 23:19-38, Spring 1990; also in *Pulse of the Planet, 3:3-16*, 1991, and *World Futures,* 30(4):247-271, March-May, 1991; Japanese Translation, *Bioenergy,* 13:2-13, 1990; German Translation, *Emotion,* 10:106-136, 1992.

James DeMeo: "CORE Progress Report #25: Two Year Research Summary — The American West, Greece, and Germany", *Journal of Orgonomy,* 25(2):181, 1991.

James DeMeo: "OROP Arizona 1989: A Cloudbusting Experiment to Bring Rains in the Desert Southwest", *Pulse of the Planet,* 3:82-92, 1991; also published as a *Special Report,* Orgone Biophysical Research Laboratory, 1991.

James DeMeo: "Research Progress Report: Desert-Greening Program, CORE Operations in the Western USA, CORE Research and Field Operations Overseas: Greece 1990, Germany 1989-1990 Orgonomisches Projekt Waldheilung, Core Breakthrough in California Rains, March 1991, Cloudbuster Icarus", *Pulse of the Planet,* 3:110-116, 1991.

James DeMeo: "Desert Expansion and Drought: Global Interconnections and the Desert/ Drought Map", Presented to the Annual Meeting, *Association for Arid Lands Studies,* Portland, Oregon, April 1990, Abstracts of Papers, p.40.

James DeMeo: "Preliminary Results of a Desert Greening Experiment in the American Southwest, Summer 1989", Presented to the *Association for Arid Lands Studies,* Annual Conference, Portland, Oregon, April 1990. Updated and also presented to the Annual Meeting, *Association of American Geographers,* Washington, DC, 1991, Abstracts of Papers, p.46.

James DeMeo: "CORE Progress Report #26: California Drought of 1990-1991, Part II. With a Special Note on Underground Nuclear Testing and the Oakland Wildfires", *Journal of Orgonomy,* 26(1):,49-71 1992; parts reprinted in: "The Oakland Wildfires of October 1991", *Pulse of the Planet* 4:117-118, 1993.

James DeMeo: "CORE Report #30: The Desert Greening Project in Israel 1991-1992", *Journal of Orgonomy,* 26(2):248-265, 1992.

James DeMeo: "Climatic Causation and Pre-Columbian Contacts in Global Patterns of Human Behavior", Presented to the *Association of American Geographers,* Annual Meeting, San Diego, 1992, Abstracts of Papers, p.54-55. Also published as a chapter in *Saharasia,* 1998.

James DeMeo, Editor: *On Wilhelm Reich And Orgonomy (Pulse of the Planet #4),* Natural Energy Works, Ashland, Oregon 1993.

James DeMeo: "OROP Israel 1991-1992: A Cloudbusting Experiment to Restore Wintertime Rains to Israel and the Eastern Mediterranean During an Extended Period of Drought", *Pulse of the Planet* 4:92-98, 1993.

James DeMeo: "Research Reports and Observations: Orgonomic Project Waldheilung 1989 - 1993, OROP Namibia 1992-1993, Cloudbusting is Not 'Weather Modification'", *Pulse of the Planet* 4:114-116, 1993.

James DeMeo: "Weather Anomalies and Nuclear Testing: The Oakland Wildfires of October 1991", *Pulse of the Planet* 4:117-120, 1993.

James DeMeo: "Similarities and Differences Between Wilhelm Reich's Orgone Accumulator and Giorgio Piccardi's Shielded Enclosures", Presented (in absentia) to the *International Symposium on Relations of Biological and Physicochemical Processes with Solar Activity and Other Environmental Factors*, Pushchino, Russia, 1993, Abstracts of Papers, p.229.

James DeMeo: "The Desert-Drought Map and its Implications", Presented to the *90th Annual Meeting, Association of American Geographers,* San Francisco, California, 29 March - 2 April, 1994, Abstracts of Papers, p.81.

James DeMeo: "Energy in Space: Empirical Evidence and Implications for Orthodox Theory",º Presented to the *Pacific Division, American Association for the Advancement of Science,* Annual Meeting, June 1994, Special session on *"Challenges to Contemporary Views in Physics and Astronomy",* San Francisco State Univ., California, Proceedings 13(1):64.

James DeMeo: "Desert-Greening Cloudbusting Experiments in Israel, Namibia and the Horn of Africa: 1991-1994," Presented to the Annual Meeting of the *Association of Arid Lands Studies,* Western Social Science Association, Oakland, California, Session on "Climate, Climatic Change and Environmental Quality", April 1995, Abstracts of Papers, p.21.

James DeMeo: "Global Desert Haze/Dust Transport: An Interconnecting Common Denominator for Deserts, Droughts and El Niño?", Presented to the Annual Meeting of the *Association of Arid Lands Studies,* Western Social Science Association, Oakland, California, Session on "Climate, Climatic Change and Environmental Quality", April 1995, Abstracts of Papers, p.21.

James DeMeo: "Cloudbusting: Growing Evidence for a New Method of Ending Drought and Greening Deserts", *AIBC Newsletter,* American Institute of Biomedical Climatology, Sept. 1996, #20, p.1-4.

James DeMeo: "Dayton Miller's Discovery of the Dynamic Aether Drift", Presented to the 72nd Annual Meeting of the *Southwestern & Rocky Mountain Division, American Association for the Advancement of Science,* Northern Arizona Univ., Flagstaff, Arizona, 2-6 June 1996. Program and Abstracts, pp.41-42.

James DeMeo: "Independent Discovery of a Dynamic Bio-Cosmic Energy in Space and in the Atmosphere: Miller, Reich, Piccardi, Burr & Brown", Presented to the 72nd Annual Meeting of the *Southwestern & Rocky Mountain Division, American Association for the Advancement of Science*, Northern Arizona Univ., Flagstaff, Arizona, 2-6 June 1996. Program and Abstracts, pp.41-42.

James DeMeo & Bernd Senf, Editors: *Nach Reich: Neue Forschungen Zur Orgonomie, Sexualökonomie, Die Entdeckung der Orgonenergie,* Zweitausendeins, Frankfürt, 1997.

James DeMeo: "Practical Applications of Aether: Novel Biological and Atmospheric Experiments", Presented to the 74th Annual Meeting of the *Southwestern & Rocky Mountain Division, American Association for the Advancement of Science,* Mesa State College, Grand Junction, Colorado, 17-21 May 1998, Program and Abstracts, p.45.

James DeMeo: "Critical Review of the Shankland, et al, Analysis of Dayton Miller's Ether-Drift Experiments", Presented to the *Natural Philosophy Alliance,* Berkeley, May 2000.

James DeMeo: "Wilhelm Reich's Atmospheric Research - Empirical Confirmations", *Society for Scientific Exploration* annual meeting, La Jolla, California, June 2001, Abstract in *The Explorer: Newsletter of the SSE,* Vol.17, Winter-Spring-Summer 2001, p.8.

James DeMeo, Editor: *Heretic's Notebook: Emotions, Protocells, Ether-Drift and Cosmic Life-Energy, with New Research Supporting Wilhelm Reich* (*Pulse of the Planet* #5, Orgone Biophysical Research Lab, Ashland, Oregon 2002.

James DeMeo: "Dayton Miller's Ether-Drift Experiments: A Fresh Look", *Pulse of the Planet,* 5:114-130, 2002. Internet posted: http://www.orgonelab.org/miller.htm

James DeMeo: "Orgone Accumulator Stimulation of Sprouting Mung Beans", *Pulse of the Planet,* 5:168-176, 2002.

James DeMeo: "Green Sea Eritrea: A 5-Year Desert Greening CORE Project in the SE African Sahel", *Pulse of the Planet,* 5:183-211, 2002.

James DeMeo: "Origin of the Tropic Easterlies: An Orgone-Energetic Perspective", *Pulse of the Planet,* 5:212-218, 2002.

James DeMeo: "The Earth-Atmosphere Electrical Potential as a Possible Source of Energy", *Pulse of the Planet,* 5:225-226, 2002.

James DeMeo: "A Dynamic and Substantive Cosmological Ether", *Proceedings of the Natural Philosophy Alliance,* Cynthia Whitney, Editor, Vol.1, No.1, Spring 2004. Arlington, MA. p.15-20. Internet Publication: http://www.orgonelab.org/DynamicEther.pdf

James DeMeo: "New Observations on Spontaneous Heat Production Inside the Reich Orgone Energy Accumulator", Abstracts of Papers booklet, *New Research in Orgonomy Conference,* Greensprings Center, Ashland, Oregon, 6-7 August 2005. Internet Posted: http://www.orgonelab.org/Abstracts2005.pdf

James DeMeo: "The Growing Nonsense About Wilhelm Reich and Orgone Energy on Global Internet", Internet Publication only: http://www.orgonelab.org/orgonenonsense.htm

James DeMeo: "Water as a Resonant Medium for Unusual External Environmental Factors", at the *Fourth Annual Conference on the Physics, Chemistry and Biology of Water,* Mt. Snow, Vermont, October. Abstract published in: *Water: A Multidisciplinary Research Journal,* Vol.2, Supplement 1.

A NOTE OF CAUTION TO THE EXPERIMENTER

The present work discusses a new method by which natural atmospheric energetic pulsation and rainfall can be restored to droughty and also semi-arid desert landscapes. They are being made available to the public for reasons of historical and scientific documentation only, and should not be taken as an act of encouragement to anyone who wishes to build or use a cloudbusting apparatus in an unstudied or untrained manner. Use of a cloudbuster in untrained hands, or by people with pathology of character, poses significant risk to both the individual operator's health, and to the natural ecology and community in which it is used. While proper use of a cloudbuster may terminate a drought, bring gentle rains, or reduce the energy level of a dangerous storm cloud, its improper use in untrained, unskilled hands may trigger or intensify a drought situation, or create major flooding or a tornado disaster. The apparatus itself is merely a tool, which has the potential for doing great good or harm. The most centrally-important aspect of Reich's discovery on the cloudbuster is not the instrument itself, but in the operational techniques, in knowledge of the life-energy, its properties and behavior, and how it functions in relationship to what empirical atmospheric science has detailed. It is therefore more like a form of *atmospheric acupuncture*, where the cloudbuster "needle" is far less important than knowledge of the life-energetic currents and conditions in the atmosphere, and how that relates to weather.

A working cloudbuster will also accumulate a very strong charge of generally stagnated orgone energy around itself, which can be toxic. It is unsafe to stand near to, or handle the apparatus during operations. The cloudbuster operator, and any nearby casual observers, are at risk for mild to serious illness from the field effects of the apparatus. Severe "flu" symptoms, or headaches and nausea may be experienced by operators during or after working with the device, and I personally know of cases where serious skin disorders, strokes and even cancer tumors were triggered within days of cloudbuster exposure. We are dealing with a powerful current of pulsating atmospheric energy, which, as concentrated around a working cloudbuster, can expose the operator to a potential hazard equal to what was experienced by early researchers in atomic radiation, who often paid a high price with their personal health. Every serious and educated person using the cloudbuster takes a number of precautions to avoid such problems, which may potentially be deadly. Primary among these precautions is biophysical emotion-release orgone therapy, to soften and remove emotional character armor — this is a prerequisite for honest work with the device, no less than gaining accurate knowledge about it, given how biophysical armoring reduces the capacity of the operator to feel or see the atmospheric orgone energy, or to interpret the charge, flow and subjective state of the orgone during operations. But it is not just the operator who might be at risk. There are cases of severe droughts with wildfires, or massive floods and tornadoes being produced by errant cloudbusting, as undertaken by ignorant and arrogant individuals. Specific examples of these are found in my paper *"So You Want to Build a Cloudbuster"*, which is posted on-line at the website given below.

There also are major issues raised about *when* and *for what exact reasons* a cloudbuster project is undertaken, the question of it's necessity. Ending a severe drought or heat-wave, or working against wildfires are certainly good reasons, assuming one knows what they are doing. Working in ignorance of the basic fundamentals of orgone biophysics can actually worsen a drought, or inflame a wildfire, so in-depth study and training is a must. Beyond that, a serious issue has arisen in recent years on the subject of *desert-greening,* which constitutes *a potentially risky form of climate modification*, and about which the current knowledge from both orgone biophysics and classical environmental and climate science warns us against. Long-distance and possibly long-term negative effects are possible. A desert land might benefit from such a greening, but it becomes a disaster if accomplished only by displacing the dry desert air mass off to plague some other world region, placing a moist region under severe long-term drought. Several catastrophes along these lines have already been witnessed, from haphazard operations undertaken by untrained persons in Algeria, in efforts towards greening the core region of the Sahara. This makes no more sense than trying to warm up the polar regions so one might grow bananas, or make the warm rainy tropics drier and cooler. Evidence from ancient climates suggests, a wetter Sahara is found only when Europe and North America are under ice-box conditions, given how the polar jet streams move far to the south. Could "greening the Sahara" bring about this same result today? For these and other reasons, international regulation and oversight of cloudbusting activity will eventually become necessary.

In short, the cloudbuster is not a toy for the merely curious, nor is its operation as simple as its construction would suggest. If the reader is casually thinking of building one and "trying it out", layperson, Ph.D., or whatever, my advice is, simply, **Don't**. Cloudbusting requires as much time, preparatory study and work, dedication of purpose, and skill as is required for other professions of great responsibility, such as piloting a passenger jetliner, or large ocean-navigating ship, or doing delicate surgery. The fact that cloudbusting and orgone biophysics have been ignored or denigrated by many scientists working in traditional research institutions does not change this fact in the least. If your area is suffering from a drought, and you wish to have a cloudbuster brought into your area to bring atmospheric relief, this can be arranged. There also are various research seminars focused upon orgone biophysics, offered periodically across the USA, providing a means by which the serious student or scientist may learn about Wilhelm Reich's work, and about orgonomy, the science of the life energy, from which cloudbusting developed. And finally, for the advanced and serious student, apprenticeship in the actual techniques of *Cosmic Orgone Engineering* is a possibility.

More information:
* *So, You Want to Build a Cloudbuster?* www.orgonelab.org/sobuildaclb.htm

ACKNOWLEDGMENTS (1979)

This work was the initial exploration of a topic which, perhaps, could not have been undertaken at any university were it not for the unique set of circumstances and personalities involved. A general note of thanks goes from me to the whole Department of Geography/Meteorology at K.U. for being open-minded enough to allow, and in some cases even encourage, research on this controversial subject, particularly to my committee members, Professors Nunley, Augelli and Eagleman.

I am deeply indebted to Robert Morris who, for the years prior to my arrival in Kansas, gave me needed encouragement to follow my interests, and provided valuable ideas for the structuring of both the experiments and the thesis itself. A special note of thanks also to my friend Don Bill, who assisted greatly in construction of the cloudbuster, "Icarus", used in this study. I am also grateful to Professor Robert Haralick who initially introduced me to the possibilities for study at K.U. I am especially grateful to Professor Robert Nunley, who not only served as adviser and teacher, but made his farm available to me for field testing of the cloudbuster. Dr. Nunley also made valuable suggestions throughout the experimental stages of this work, and provided critical review of the thesis in his role as chairman. Professor Joe Eagleman also supplied many valuable criticisms of the thesis, as well as having served as a constant source of information during the work stages. Professor John Augelli also provided needed review of structure, logic and clarity of a topic admittedly difficult to approach. My thanks also to the meteorological staff of the National Weather Service offices in Topeka, Kansas, who provided access to satellite imagery, to Mark Bernstein, who provided ideas on photography, and to Dr. Curtis Conley, of the Kansas Geological Survey, who provided generous use of their VP-8 image analyzer. Thanks also to my father, Vincent DeMeo, who financially assisted me in this study, and who has always been a source of enthusiastic encouragement.

This thesis was typed and printed on a CPT automated text-editing system generously made available by Dr. Robert Nunley and Child, Inc. of Lawrence, KS

PRELIMINARY ANALYSIS OF CHANGES IN KANSAS WEATHER
COINCIDENTAL TO EXPERIMENTAL OPERATIONS
WITH A REICH CLOUDBUSTER

by

V. James DeMeo, Jr.
B.S., Florida International University, 1975

Submitted to the Department of Geography and the Faculty of the Graduate School of the University of Kansas in partial fulfillment of the requirements for the degree of Master of Arts.

Robert E. Nunley
Chairman

John P. Augelli

Joe R. Eagleman

CHAPTER I

INTRODUCTION

Current meteorological theory ascribes the phenomena of weather to various mechanical-energetic forces, the most successful model of which is the view of the atmosphere as a giant, solar-driven heat engine. This model embodies the following basic factors:

1. Incoming solar radiation is distributed unequally between the equator and poles.

2. Energy absorption and transference to atmosphere at the earth-atmosphere interface is different between crustal and ocean surfaces.

3. Diurnal changes and seasonal cycles occur in solar-earth configurations causing time and location variations in terrestrial energy buildup.

4. The atmosphere and oceans are gravity driven to cycle heat energy, manifest as density-weight differences, between equator and poles.

5. Energy is absorbed via water evaporation and released into the atmosphere, generally at higher altitudes and at higher latitudes, via precipitation, latent-heat processes, and via direct transfer of heated air masses.

6. Inertial and thermal exchange occurs in net fashion from the equator toward the poles.

7. Air masses possess certain physico-chemical properties which change as they move across the terrestrial surface; such changes determine the interaction of the air mass with respect to water and thermal energy.

8. Air masses of different properties interact, as the properties of a single air mass may be altered, to cause the phenomena of weather.

9. Absorbed solar radiation is eventually dissipated to space as long wave radiation.

However successful this solar heat-engine model may be in explaining most atmospheric phenomena, there are many occurrences in the natural world involving the atmosphere which defy easy theoretical integration. Current scientific discovery, largely via the space program, has provided an enormous element of mystery to our understanding of the atmosphere, the earth, the sun, and space itself. The role of the ionosphere, charge separation in thunderstorms, the earth's electrical field, magnetosphere and radiation belts, as well as sunspot, solar flare and sector-boundary relations to earthly weather are scarcely understood, much less incorporated into the daily activity of the weather forecaster.

We note that no notion of continuity or interaction between separate locations within the atmospheric ocean is generally accepted other than that which could occur solely via transference and mixing of the different air parcels, or by some other thermo-mechanical or inertial transfer, such as jet-stream interaction. Electrical and electromagnetic phenomena occurring within the atmospheric ocean are largely seen as mere by-products of thermo-mechanically built-up weather, rather than as a manifestation of any kind of primary causal or triggering agent. Ideas on electrical influences in the atmosphere are also limited largely to micro-meteorological parameters, such as charge on cloud droplets. There is currently no agreement on the existence of any principle of continuity in the atmosphere which would allow an event at location X to affect or trigger a reaction at some distant location Y, other than via the mechanical pathways outlined above.

The present study will cover the result of two years research with a weather modification technique that deals directly with this question of atmospheric continuity. The technique is called “cloudbusting” and the device used is called a “cloudbuster”. The simplicity of the cloudbuster’s design will naturally raise theoretical difficulties as to its possible influence upon the atmospheric ocean, particularly among those who would argue against the existence of a principle of atmospheric continuity. However, we may note the development of many other simple devices which, in their time, via interaction with a poorly-understood medium, produced an amazing result. Examples are the lens (medium = light), the lightning rod (medium = electrons), and the airfoil (medium = air). Should the cloudbuster prove to be an operational device, classical meteorological theory would be hard pressed to say just what the “medium” involved is.

In order to disencumber the present study of theoretical objections, we make the assumption that such a principle of atmospheric continuity does exist. We leave absolute proof of, and elaboration upon, such a principle for later study, noting that classical evidences supporting this assumption have been gathered in Appendix I. The task in the present study is to experimentally evaluate the cloudbuster device, allowing any empirical evidence gathered to define our position.

Statement of Problem

Reduced to its simplest form, a cloudbuster consists of a parallel array of hollow metal tubes, about 18 feet long, extending up from a clean, running water source. The tube array is aimed at different areas of the sky, or to distant locations, for different purposes. It remains on the ground during the entire operation. The cloudbuster is a passive device, working to trigger movements of the atmosphere by virtue of the pre-existing energetic properties of the atmosphere itself. In this manner, a cloudbuster is much like a lightning rod. However, unlike a lightning rod, the cloudbuster is grounded into a source of pure water, preferably running and pollution-free, via hollow metal tubes, rather than to earth-ground via conductive cables. No chemical inoculants are used, as with cloudseeding; neither is electrical or electromagnetic energy actively provided to the machine, other than 12 volt D.C. to run relays and servo motors for aiming the heavy array of metal tubes. Figure 1 is a photograph of the cloudbuster used in the present study, a full description of which appears in Appendix II.

Reich (1952), who invented the cloudbuster, describes its operation as occurring through manipulation of a ubiquitous, heretofore undiscovered energy continuum existent in the atmosphere. This energy was named the orgone (Reich 1942, 1948) as it was first discovered to exist in living organisms. Reich describes the orgone as a continuum which charges and penetrates all substance, living and non-living. It is also deemed responsible, depending upon its concentration and flow in the atmosphere, for the phenomena of weather. Temperature, pressure, humidity, etc. are considered, in the framework of orgone theory, as secondary manifestations of activity within this underlying energetic continuum; modification of properties of this energy continuum are said to cause change in meteorological parameters. Appendix III gives a detailed summary of literature on Reich's orgone theory, the "proving" of which is not our task here.

Reich describes the orgone charge of a cloud as the cohesive force which holds it together against the forces of diffusion and mixing: the orgone is described as possessing both a negative entropy and a mutual affinity for water. The hollow tubes of the cloudbuster focus on and "draw" energy from specific regions of the atmosphere, grounding it into a water source the other end of the tubes are immersed in. The induced energetic movement is said to trigger movements in the target weather system, which are considered to be energy dependent. We should note that Reich's orgone theory is empirically based using traditional scientific methodology.

The cloudbuster is a simple device, so much so we may honestly ask how such an arrangement of materials could have any effect whatsoever. In this regard, the Department of Commerce, which licenses weather modification activities within the U.S.A., has stated that cloudbusting is "not acceptable from a technical or scientific viewpoint".[1] However, they also state that their "negative opinion is based on theoretical considerations, not tests".[2] They did not say just what these theoretical considerations were but did request that any additional evidence be sent to them. Cloudbusting operations of about a half dozen individuals in the U.S.A. are filed under the Department of Commerce reporting plan, and such activities are listed in their Summary Reports (U.S.D.O.C. 1974). Even though meteorologists in general are unaware of the cloudbuster, or reject it on theoretical grounds, a large body of positive empirical evidence has been accumulating. Basic principles of cloudbusting are given by Reich (1952) and reiterated by Kelly (1961). Aiming the cloudbuster at moderate sized cumulus clouds was found to trigger their dissipation; aiming at open sky adjacent to existing cumulus increased their growth. Such effects are noted as commencing almost immediately.

Applications of the cloudbuster for drought relief or "rainmaking" are reported by Reich (1952, 1954, 1957), Moise (1955), Eden (1972, 1973), Constable (1971, 1972), and Blasband (1970, also see "CORE Progress Reports" 1970 - 1979). A technique for fog-lifting is reported by Reich (1954) as well as a theoretical technique for engulfing a wide area under a fog blanket. McCullogh (1955) has explored experimentally the relationship between cloudbusting and cloudseeding.

Reich (1954, 1955a, 1957) led an expedition into the desert near Tucson, Arizona for the expressed purpose of bringing rain. He claimed to have been successful, temporarily at least, in using a series of special cloudbusting techniques for drawing in moisture from the ocean to the west. Reich also expounded on ways in which, using strategically positioned cloudbusters, precipitation in broad

---

1. Letter to James DeMeo from R. Lavoie, Environmental Modification Office, D.O.C., 25 January 1977

2. Letter, ibid, 12 April 1978

areas of the American Southwest could be enhanced. The problems of desert, relative to cloudbusting, have also been touched by Silvert (1955) and Blasband (1970).

Wind direction, cloud patterns and large scale storm systems reportedly respond to the cloudbuster's triggering influence (Reich 1952; Grossman 1955; Kelly 1955; McCullogh 1955; Blasband 1969). Two reports are also given of hurricane diversion, by Reich (1954) regarding hurricane Edna, and Blasband (1972) regarding hurricane Doria.

Blasband (1970) has also related results of a series of cloudbuster rainmaking operations:

> "Our criteria of success for a rainmaking operation is as follows:
> 1. Forecast of rain probability is 10% or less for the next 24 hours, and 'no' precipitation is expected for 48 hours. (In practice no rain was usually in sight for at least three to four days.) 2. It actually rains, even if only a trace, within 48 hours of the start of the operation...
> In 1965, we found that, out of 38 operations to engineer rain under the above criteria, we were successful 18 times. After allowing for a 30% chance of mistaken forecast, we still find that this is 1.6 times more rainfalls than one would have expected by chance alone. Distributed throughout the droughty part of the year, these seven additional rainfalls brought considerable relief."[3]

The Israeli meteorologist Assaf has similarly related results of cloudbuster tests during which, in 1963 - 1964, the "500 mb surface descended by 100 meters or more, as compared with the objective barotropic prediction". Assaf concluded that the effect of his tests stretched 1000 - 2000 kilometers in the direction of the draw.[4]

Most of the individuals who have worked with the cloudbuster have done so in a quiet manner and published in journals generally unknown to meteorologists. None were connected with the university system or was supported by public funds in their work. Also, unjustified repression of Reich's ideas by the scientific establishment was substantial (Greenfield 1974; Wolfe 1948). The Pure Food and Drug Administration, acting on a scanty amount of hastily set-up and poorly controlled experiments, obtained an injunction against the continuation of Reich's research work and against distribution and sale of any book or journal bearing the word "orgone". Reich's research laboratory in Rangeley, Maine, was

---

3. Blasband 1970, p. 68 - 69

4. Letter to James DeMeo from Gad Assaf, 12 February 1979

invaded by agents of the F.D.A., who put his scientific instruments to the axe; books and journals were carried off to New York incinerators in 1957 and later in 1961. Reich was later found to be in technical violation of the injunction and was sentenced to two years in Lewisburg Federal Penitentiary. He died two weeks prior to his release date. Rebuttal to the F.D.A. "evidence" against Reich has been undertaken elsewhere (Blasband & Rosenblum 1972), and since the F.D.A. did not involve itself with any of Reich's atmospheric work specifically, we shall content ourselves here with a mention of the difficulty.

We therefore have a situation wherein a scientist of world renown in the field of medicine makes claim of a discovery which is largely ignored by his contemporaries, in and out of medicine. His work eventually takes him from biomedical to atmospheric considerations. He is later thrown into jail, his work supposedly having been "disproved" by a large government bureaucracy. Yet, as we have shown, his coworkers continue to publish results very positive to the man's discovery. And, upon a close examination, we find the bureaucratic "rebuttal" to have been incomplete, poorly controlled, and perhaps even politically motivated.[5] Hence, we are left with nothing but positive evidence for Reich's experimental works. And, hence, the need for scientists to address the problems and possibilities in a responsible fashion.

We note that documentation of any weather modification technique is a difficult task. Even the technique of cloudseeding still has its disbelievers in the scientific establishment. Claims for the effectiveness of cloudseeding range from 5% for an increase in precipitation (Eden 1978) to over *510%* (U.S.D.G.C. 1972).[6]

Cloudseeders have certainly shown their technique to have an effect upon the atmosphere. However, they may be either underestimating or overestimating this effect. This is so even though cloudseeders have enjoyed multi-million dollar support and years in which to pin down the exact nature of their work. In contrast, cloudbuster research, including this study, has been supported wholly out-of-pocket. Hence, it is hoped that the reader will keep in mind the preliminary nature of this study and not unduly burden it with criticisms that could only be answered with multimillion dollar support.

---

5. Greenfield 1974; Wolfe 1948

6. Some have even laid responsibility for "massive, destructive and widespread" floods and general weather chaos at the doorstep of the cloudseeders based on alternative meteorological theories which closely mirror our "principle of atmospheric continuity". (cf ACRES, USA 1977, Langmuir 1961)

Experimental Design

In spite of the above obstacles, a respectable cloudbusting apparatus was constructed and tested. Data from the National Weather Service was used to document change in weather over the state of Kansas coincidental to operations of the device. Three basic tests were made with the cloudbuster which will be related in detail shortly:

1. Tests for cloud dissipation

2. Tests for precipitation abatement

3. Tests for precipitation enhancement

Results of these tests, which ran from June 1977 through October 1978, will be presented in an historical manner. Tests on cloud dissipation were first undertaken, followed by intermittent tests for precipitation abatement and enhancement. Each chapter relates problems specific to the particular test involved.

It should be noted that these tests were initially begun without any long term experimental plans other than to make repeated general observations. These tests were also the trial operations of a newly constructed cloudbuster which the operator, myself, was slowly becoming familiar with. Mechanical debugging of the device also occurred during this period. I had occasionally worked with other cloudbusting devices, mostly smaller ones, over the previous seven years.

Tests for cloud dissipation were eventually worked out into a controllable technique that could be run on a given day. However, tests for precipitation abatement and enhancement presented special problems of control and design. For instance, the earliest use of the device was based on a subjective evaluation of environmental "need". If local flooding seemed likely, the precipitation abatement test was run; if a prolonged dry spell occurred, a precipitation enhancement operation was performed. Consequently, one will note that all three of the precipitation abatement operations and four of the 12 precipitation enhancement operations were done on a random basis. The day and time of these operations was determined by my "feel" for the need and by local forecasts which generally indicated a low probability for change in the weather.

By July 1978, it was decided to undertake a series of pre-planned precipitation enhancement tests. This was done primarily to overcome any possibility that I might have been unconsciously selecting for conditions favorable for rain. These pre-planned operations, which occurred on successive Sundays, are described in detail in the section on Tests for Precipitation Enhancement.

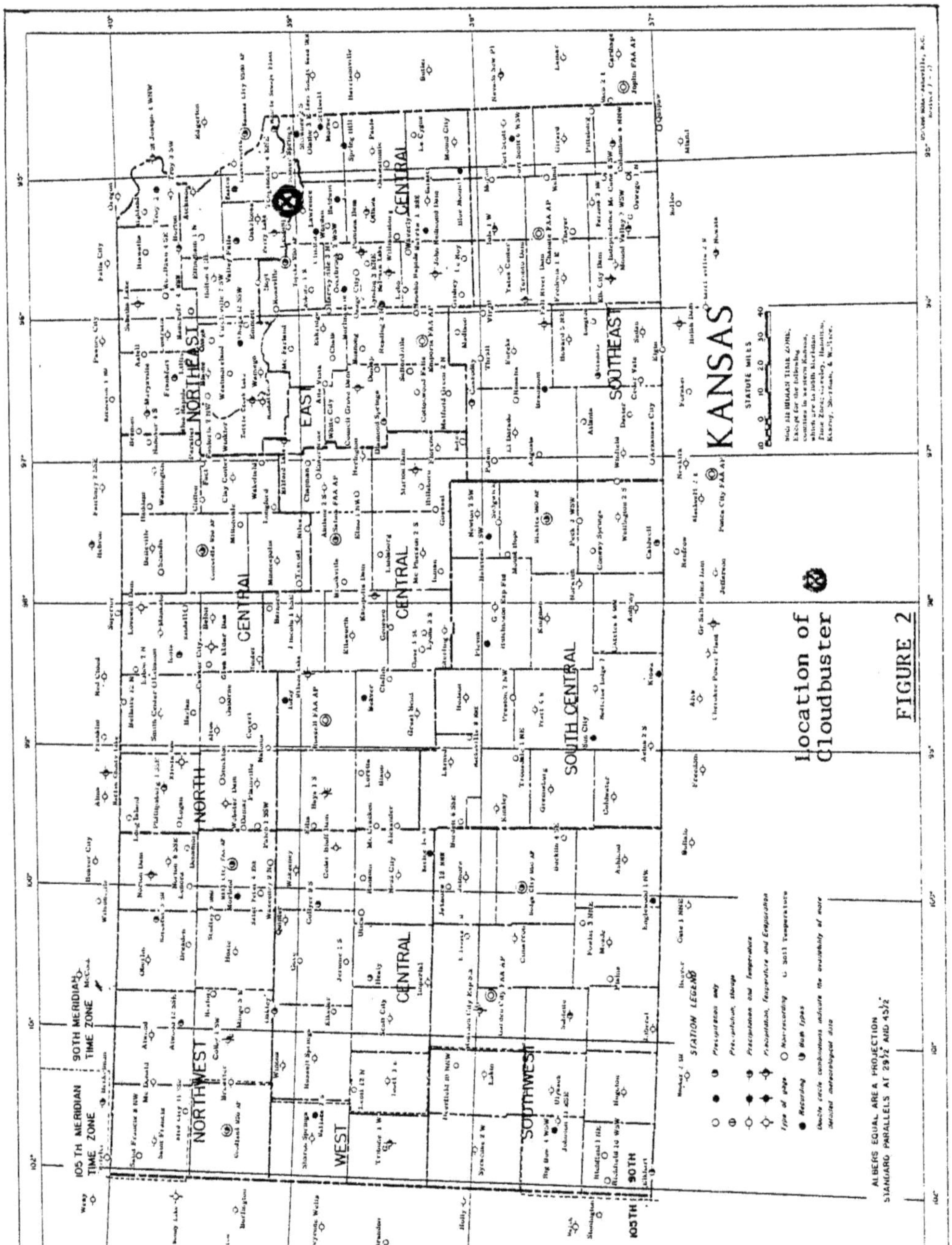

Location of Cloudbuster

FIGURE 2

In each chapter is found a section relating general qualitative observations, many of which were not documented during these trials. These observations showed up with enough regularity so as to make their causal association with the cloudbuster seem probable. This qualitative material is followed in each chapter by quantitative, documented observation, using data gathered at the test site or obtained from National Weather Service records. Hence, the most convincing evidence appears toward the end of each chapter. And, each of the three chapters on cloudbuster tests are, I feel, successively more convincing.

Notes on Data and Study Area

All the tests related in the present study were done at a field site about 10 miles north of Lawrence, Kansas. Figure 2 is a map of Kansas showing the location of the field site and weather stations from which data were taken. All data presented in Chapter II, on Tests for Cloud Dissipation, were gathered by myself at the test site during operations. Data for Chapters III and IV, on tests for Precipitation Abatement and Precipitation Enhancement, were mostly taken from the following Department of Commerce publications:[7]

1. Hourly Precipitation Data, from 73 Kansas weather stations, is published about 9 months after records were initially taken. Hence, that material is used for hourly precipitation measurements only up through May 1978.

2. Local Climatological Data, from five Kansas stations and one in Missouri (Kansas City), were used for the hourly precipitation summaries after May 1978. Percent cloud cover data were also taken from this publication.

3. Climatological Data, from 278 Kansas stations, was used here for daily precipitation data. However, these data are recorded by each station at different times of day. A check of the station records indicates that 5.5% of these stations recorded data at midnight, 88.4% record data at 7 am + or - one hour, while 6.1% record data at 6 pm + or - one hour. The midnight recorded data is 7 hours off from the 7 am data, but the 6 pm data is 13 hours out of sync, since all data is recorded for the day the measurement was taken on. These data are most heavily biased (88.4%) for the 7 am observations and are valuable in this study only as long as the time differences of observations are kept in mind. While these time differences place an error in totalized data for daily precipitation values from all 278 stations combined, they do not appear to be large enough an error to prevent careful use of the data in this preliminary study. Climatological Data is published about two months after data is recorded.

---

7. U.S.D.O.C., Records Center, Asheville, N.C.

4. Storm Data, for the state of Kansas, gives dates and times for all severe storm events (tornadoes, funnel clouds, severe thunderstorms, hail, storm winds, floods, etc.). These data, which were used in Chapter V, on Cloudbuster Operations and Severe Storms, are published about 3 months after the fact.

5. Daily Weather Maps, for North America, are included in Appendix VI, for a four day period spanning each cloudbuster precipitation abatement or enhancement test.

For our study area, we use the entire state of Kansas. The state wide area was chosen, in part, due to easy availability of data, and because the scale of expected cloudbuster effects was at least that noted in literature on cloudbusting.[8] However, the question is again raised as to how such a large area could be affected by operation of a single device located in the Northeast corner of the state? Again, according to best available evidence, the device would function according to some non-mechanical parameter allowing energetic connectivity between disparate locations of the atmosphere.

The extent or degree of affect of the cloudbuster would largely be determined by the mechanism of its interaction with the atmosphere. Cloudbusting literature, which largely follows Reich's lead in using the orgone energy continuum as the mechanism, leads one to expect effects up to 2000 km from the test site. Such effects are possible within the framework of Reich's orgone theory. However, the consideration of electrical interaction between the cloudbuster device and the atmosphere has yet to be fully explored. But it is known that the protective radius of the lightning rod, a similar device, is only about two times its erected height above ground level (Chalmers 1967). The cloudbuster used in this study stands about 20 feet above ground level at maximum and was not strongly connected to earth ground during operations. Assuming a possible electrical parameter for the cloudbuster's effect, over how large an area would these effects manifest? Lightning rod theory would suggest an area about the size of a football field. Kansas is an area which is roughly in between these two extremes.

Also, since storms track generally from west to east, the state of Kansas was a properly delineated area for observation of effect upon approaching air masses. The effect of the cloudbuster on distant fronts or storm systems to the west or northwest resonates with other unanswered meteorological questions, such as non-thermal solar phenomena that associates with changes in synoptic and mesoscale circulation and precipitation patterns, or the biological phenomena which non-mechanically fluctuates with or predicts weather (see Appendix I).

---

8. I originally wanted to evaluate the four-state area of Kansas, Nebraska, Iowa and Missouri. However, this was too large of a task for this preliminary study; it will hopefully be accomplished sometime in the future.

Are such phenomena purely coincidental? Or are they evidence of a none-too-well understood principle which underlays and interconnects both earth, atmosphere, and living system? These are intriguing questions which are interwoven with the tests here, but which shall have to wait for later clarification. Certainly, to point a cloudbuster at a cloud one mile away and perceive an effect is no less a theoretical enigma than to perceive effects at 10, 100, or 1000 miles distant. In each case thermodynamic transfer and air mixing is impossible, given the time frame involved. We must, again, defer the question of mechanism for later study, and refer the reader to Appendix I for further justification of selection of the wide study area.

CHAPTER II

TESTS FOR CLOUD DISSIPATION

According to Macdonald (1958), the most meaningful way statistically to test for the effect of cloud dissipation would be to:

1. Determine the type of cloud one wishes to modify (in this case isolated cumulus) as well as other environmental parameters, then

2. Flip a coin after such a cloud comes into view; heads you test, tails you don't test, then

3. Observe both tested and untested groups of clouds for changes that develop. This gives you an identical control group for comparison with the tested group.

A Problem with the above approach is that, due in part to the dynamic nature of cloud behavior, a large number of clouds must be observed, numbering in the hundreds, to obtain statistical significance. Aside from the environmental disturbance one would probably cause with such a large number of tests, the time and resources available to this researcher, who worked alone for the most part, were extremely limited. After two years of experimental testing and debugging of the apparatus involved, I had documented only 20 individual cloud sequences which could be meaningfully compared. However, this small number of tests does not prevent preliminary types of interesting data analysis.

In this chapter we shall first relate some general observations and problems encountered. It was the working through of these early considerations which led to more specific and quantifiable tests for cloud dissipation, which are related toward the end of this chapter.

Lifetime of Cumulus Clouds

Contrary to what was expected, a reasonable search of the literature turned up no specific empirically-based work revealing the lifetime of a cumulus cloud for a given set of environmental conditions. While it is common for some meteorologists to express the opinion that a "cumulus" cloud has a "short" lifetime, usually expressed to be from 10 to 20 minutes, it is not exactly clear which type of cumulus is generally meant, nor what the expected deviation from average lifetime is. If we consider the factors involved with cumulus growth and decay, it is immediately apparent how the lifetime of such clouds could vary appreciably. Some of these factors would be humidity, temperature profile, condensation nuclei concentration, time of day, season, character of underlying surface, turbulence, etc. These factors can and do vary in different ways to develop a wide variety of behaviors in cumulus clouds.

Growth and decay in cumulus clouds are easily noted by casual observation, but deeper patterns would not be apparent if only superficial observation is maintained. Some cumulus clouds obviously dissipate within minutes, or even seconds. However, others must persist long enough to converge and grow into cumulonimbus, a process which can last for hours. My own general observations, to be related in greater detail shortly, suggest specific stages of cumulus development, each with different lifetimes.

For instance, cumulus clouds in streets were observed to be more persistent than cumulus clouds in less organized or random patterns. Isolated cumulus clouds did not show any general specific lifetime, sometimes being persistent (over 25 minutes), sometimes not. In general, cumulus in mT air masses tended to grow and decay with rapidity. This rapid pulsation would not always result in total dissipation, for just as often as not, a piece of decaying cumulus would reverse itself and blossom to full growth again. Cloud edges in mT were always more interactive with the environment, as compared with cP or mP, showing a tattered appearance (cumulus fractus). cP or mP cumulus forms were generally large, with sharper, less environmentally interactive borders, and were very durable.

General Observations

In the cloud dissipation tests run with the cloudbuster the device is first grounded into the water source, and then aimed at the center of the most massive area of the cloud with the objective of following the cloud as it moves, retaining aim on the most massive piece. Tests for cloud dissipation were run on a total of 20 different days in 1977 and 1978 (see Appendix IV for a list of all dates) with the following general qualitative observations.

1. Cumulus clouds were most affected by the device; altostratus or cirrus were least affectable in a consistent manner.

2. Cumulus on light wind days In mT ai r masses were more easily dissipated than cumulus in cP or mP air masses. (see Photo Sequences 1, 3, and 4)

3. Cumulus in streets were more resistant to dissipation than isolated cumulus.

4. Cumulonimbus showed signs of dissipation, as did nimbostratus — these effects will be dealt with in the section on Tests for Precipitation Abatement.

5. Thin layers of stratus would sometimes open up exactly where a cloudbuster was pointed, but would shortly close back up, often at a lower altitude by “scud”. (see Photo Sequence 2)

6. Large water bodies, such as lakes Pomona or Perry, worked best for demonstrating the effects of cloud dissipation. Smaller water sources, such as a 500 gallon water tank filled with running hose water, worked less well.
7. One may note that points 2 and 3 above are in general agreement with observations of unmodified cumulus under similar conditions, with respect to lifetime.

Photo Sequence 1: (on following page)

CUMULUS DISSIPATION

This sequence was taken in an mT air mass with southerly wind direction. The area of the cloud mass being aimed at dissipates while other areas of the cloud remain constant or grow somewhat. Photos were taken an 15 second intervals.

Photo Sequence 2: (on subsequent following page)

STRATUS DISSIPATION

A solid but thin stratus formation which had been stable for at least two hours prior to this test forms a large hole within a minute after the cloudbuster is aimed. This hole remained open for about five minutes before closing back up, although in a less solid manner. Photos were taken at one minute intervals.

1

2

3

4

5

Photo Sequence 1:

CUMULUS DISSIPATION

Photo Sequence 2:

STRATUS DISSIPATION

Photographic Analysis

Much time and money was wasted in attempts to catch on 16 mm time-lapse movies the cloudbuster's affect on cumulus clouds. However, problems with this technique were insurmountable for a one-man operation. On a day with good cumulus structure, assuming I was free of work responsibilities, I would drive to the field site where the cloudbuster was located. Set-up time usually took about an hour. The movie camera would be set into place ready to pan a wide region of the sky, including the projected path of an oncoming cloud and the cloudbuster itself. The sky would then be filmed in its unmodified state to establish general trends prior to modification. More than several times such a routine was completed only to have the wind change direction, making the planned panning area useless and prompting a retake later on from another orientation. At other times there would be a complete overcasting, or complete dissipation of all clouds prior to operation, which would make cloudbuster testing and filming impossible. A hand-held still picture camera was thereafter employed for documentation.

During these early trials, I often observed that some clouds would flatten out and change general direction of drift as I commenced "drawing" from them with the cloudbuster. Such flattening and "running away", as I call it, occurred only when the cloudbuster was grounded into a small water body. It reminded me of attempts to pipette up a blob of liquid mercury — the blob (cloud) would flow around and away from the tube unless you "got" it just right with your aim. This observation fits well with the notion of a "skin" phenomena postulated as binding the cloud together to overcome its expected (vs. observed) rate of diffusion and environmental entrainment.[1] With the cloudbuster, I seemed to "prick" the "skin" of the cloud.

Similarly, if I was aiming at a cloud and lost focus or aim on it, as often happened with a malfunction of directional motors, the cloud would stop its dissipation and grow again for a period. Re-aiming the cloudbuster would again start the cloud dissipating As these experimental trials progressed, it became clear that I would have to establish the following two considerations:

1. That the observed phenomena occurred in a time scale different from that expected for clouds of a given type, and

2. That the effects were repeatable for a significant number of trials, as compared to a control group. Owing to the difficulties outlined above, it was not until June 1978 that a proper test was developed to supply a matched set of control and tested clouds fulfilling these criteria.

---

1. G.A. Marotz, 1979, microclimatology lecture

On a day when the proper cumulus formations were present, the cloudbuster was set up but temporarily withheld from operation. A number of clouds were sequentially selected and photographed with color slide film at one minute intervals for around 10 to 15 minutes each. These clouds became the control group. Next, a number of clouds meeting the same descriptive criteria were sequentially photographed in their natural state for five minutes, after which the cloudbuster was pointed at them. Photographic documentation continued for another five minutes minimum, or until the cloud dissipated entirely. This second set of clouds became the tested group.

Slide film sequences of the two groups were developed and traced onto a sheet of paper. Traced cloud outlines were inked in, giving a high contrast image of each cloud for a given interval of time. These inked cloud images were then placed under a VP-8 image analyzer which gave an accurate area count of each. The assumption here is that areal changes in sequences of two-dimensional cloud images bears a close relationship to actual changes in the cloud's three-dimensional volume. (Slides could not be analyzed directly under the VP-8 due to instrumental misinterpretation of cloud background at the same grey tone level of the cloud itself.) Since each cloud was photographed for each minute of observation, we have a time series of data representative of the volume changes in the clouds.

When the cloud image data is plotted on a graph against time we obtain an interesting sequence, the significance of which depends upon 1) how strongly we value trends in the data prior to the time when the cloudbuster is aimed, and 2) the comparative values between control and tested groups, which were photographed on the same days but slightly earlier in the day. Figures 3 - 13 show graphically the growth and decay patterns of individual control group clouds for both test dates combined. Figures 14 - 22 show similar data for the cloudbuster tested group. All data points are one minute apart, as were the photographs from which they were derived.

Data presented here is representative of two days worth of filming, 30 June 1978 and 8 September 1978. photo Sequences 3 and 4 were made during these trials. A traced and inked image of all cloud sequences, for both control and tested groups, appears with their respective graphed data, in Figures 3 - 22. Cost and space limitations were prohibitive for duplicating more than a few of the original photo sequences.

Photo Sequence 3 : (on right-opposing page)

CUMULUS DISSIPATION

The cloudbuster was aimed at this cloud right after photo #3 was taken. Photographs were taken at one minute intervals. This photo sequence corresponds to the data graphed in Figure 16.

1 2 3

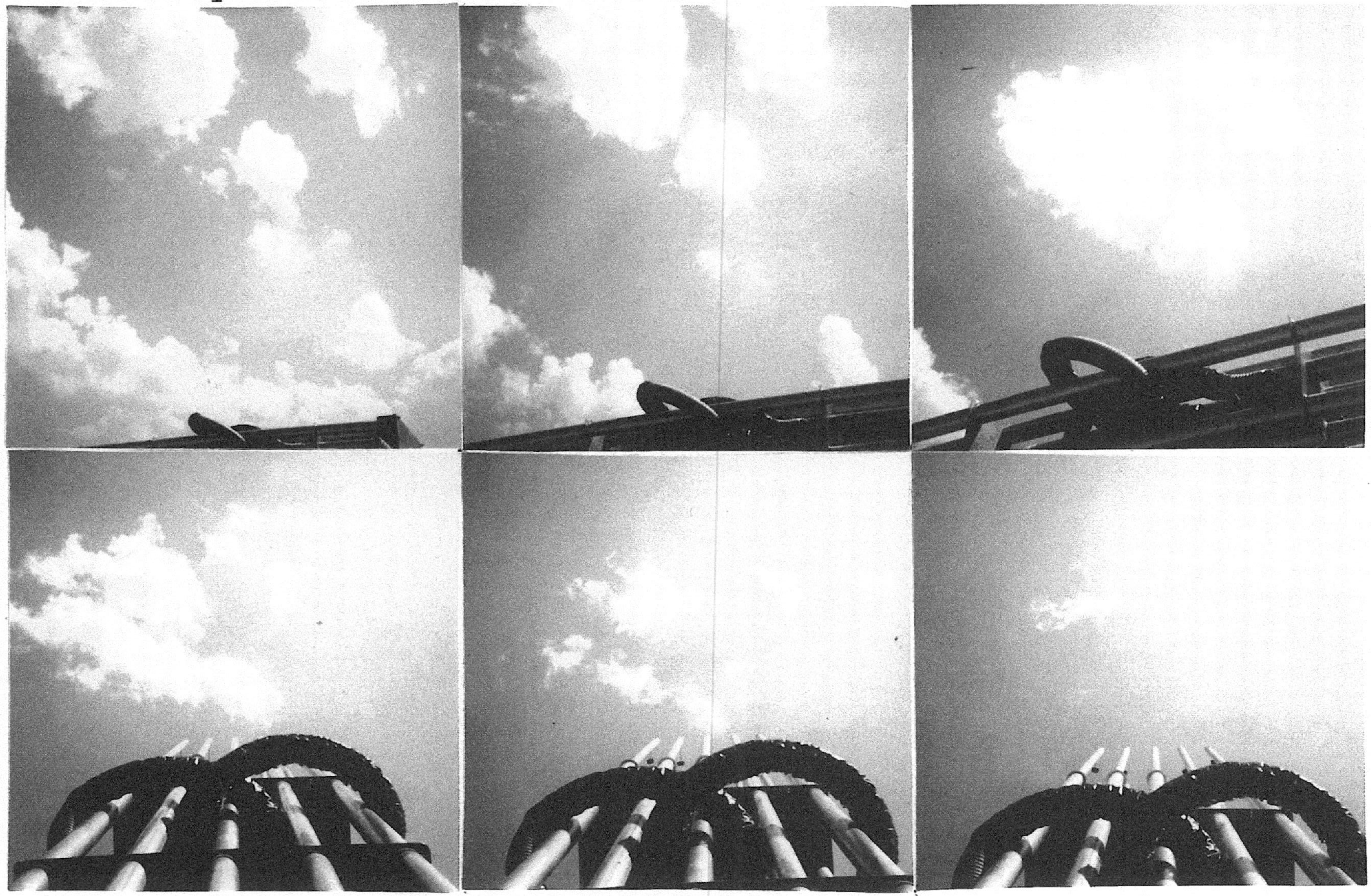

4 5 6

Photo Sequence 3: CLOUD DISSIPATION

Photo Sequence 4: (on right-opposing page)

CUMULUS DISSIPATION

The cloudbuster was aimed at the largest cumulus in this small cluster right after photo #5 was taken. Right after photo #10, the cloudbuster was re-aimed at the last remaining small cumulus in the cluster. Photos were taken at one minute intervals. This photo sequence corresponds to the data graphed in Figure 18.

Photo Sequence 4: Cloud Dissipation

Figure 3: CONTROL SEQUENCE "P": 3Ø June 1978

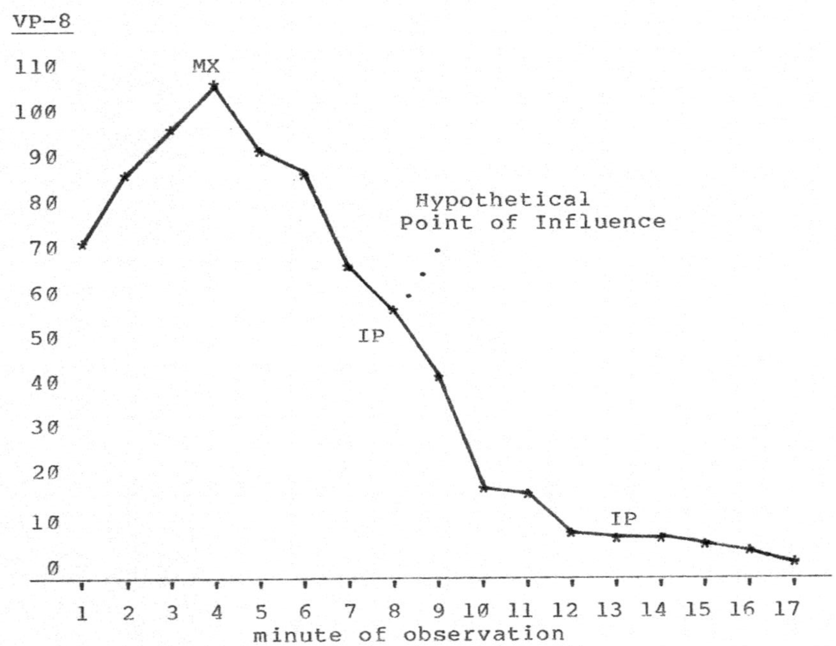

VP-8 reading:
71 86 95 1Ø7 89 87 66 57 38 17 15 6 5 5 2 1 Ø

Inked Cloud Sequences:

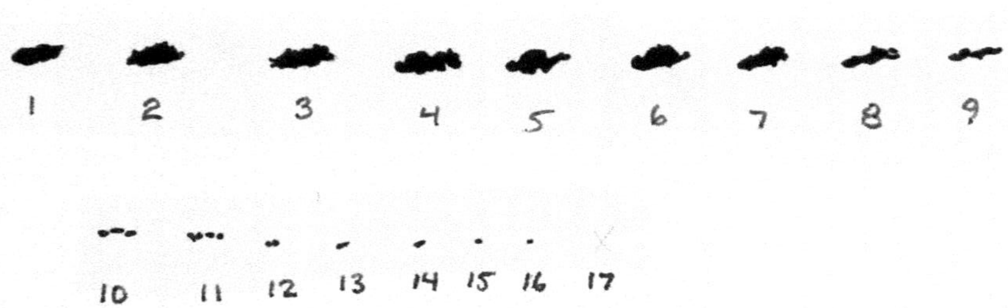

Figure 4: CONTROL SEQUENCE "Q": 3Ø June 1978

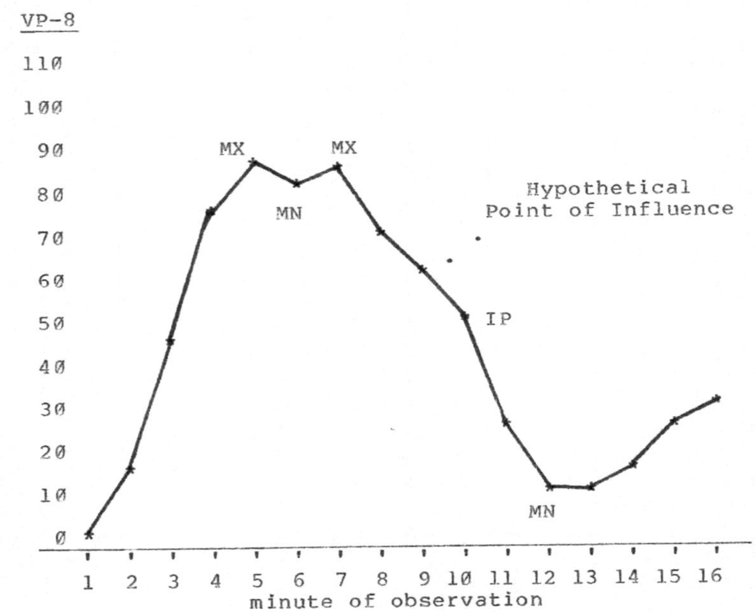

VP-8 reading:
Ø 12 44 75 87 82 85 69 62 5Ø 26 8 8 14 27 3Ø

Inked Cloud Sequences:

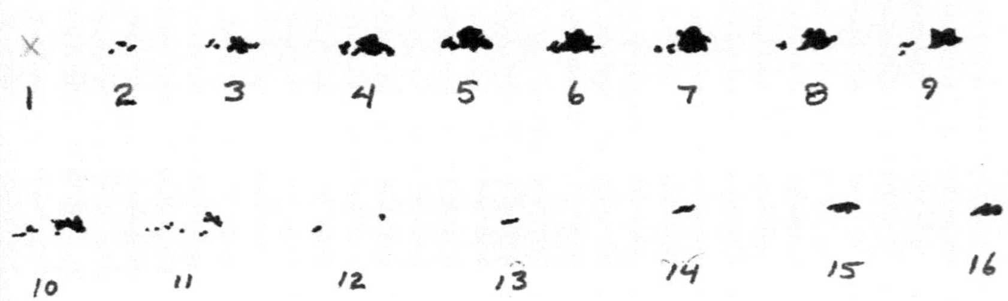

Figure 5: CONTROL SEQUENCE "R": 30 June 1978

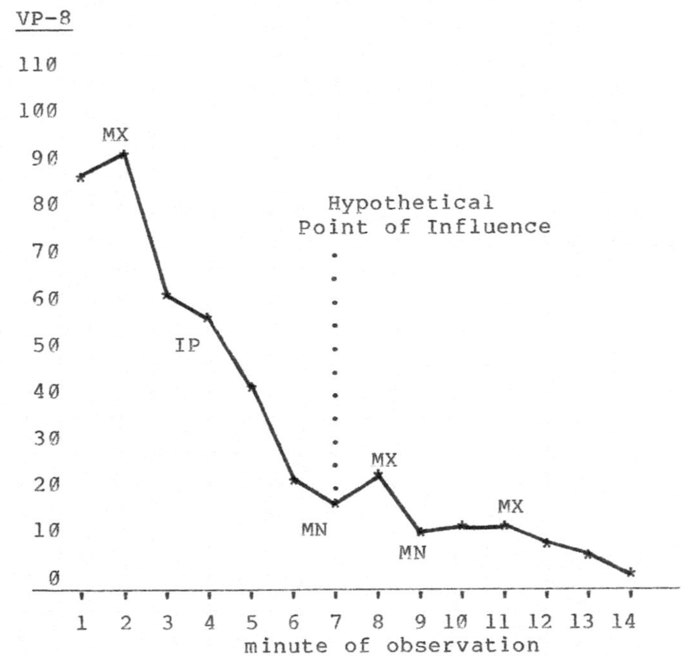

VP-8 reading:
83 89 62 53 41 19 16 21 9 10 10 7 4 0

Inked Cloud Sequences:

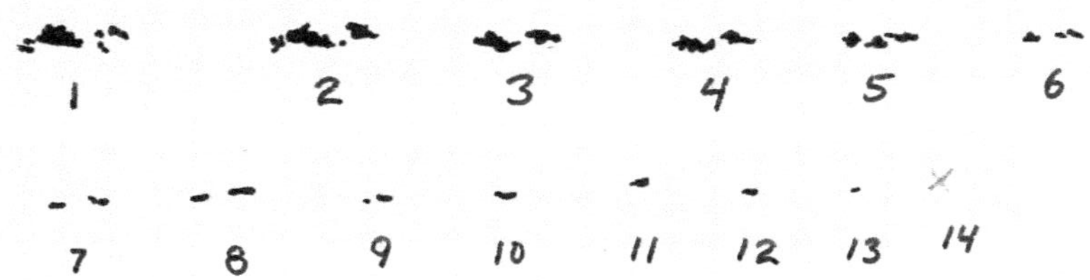

Figure 6: CONTROL SEQUENCE "S": 30 June 1978

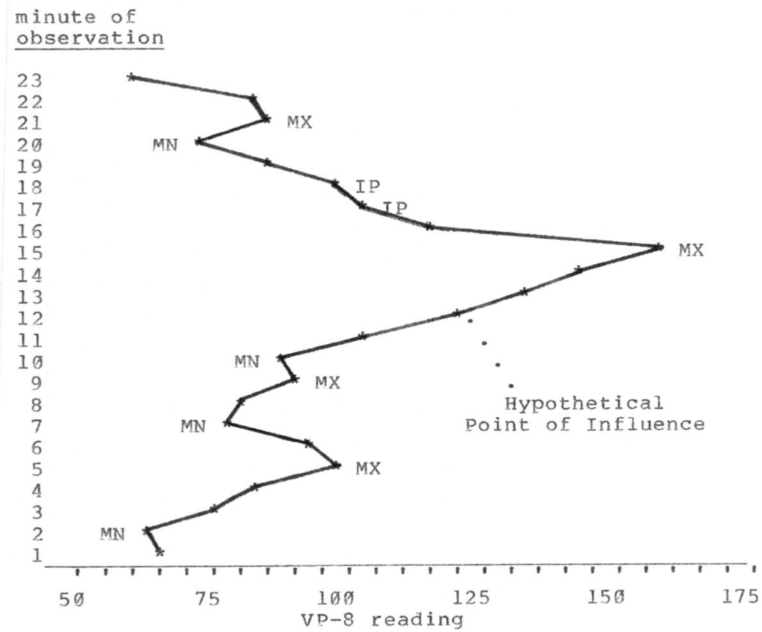

VP-8 reading:
66 63 74 82 97 93 77 79 90 89 103 121 132
142 158 115 115 101 97 83 74 84 82 60

Inked Cloud Sequences:

Figure 7: CONTROL SEQUENCE "T": 30 June 1978

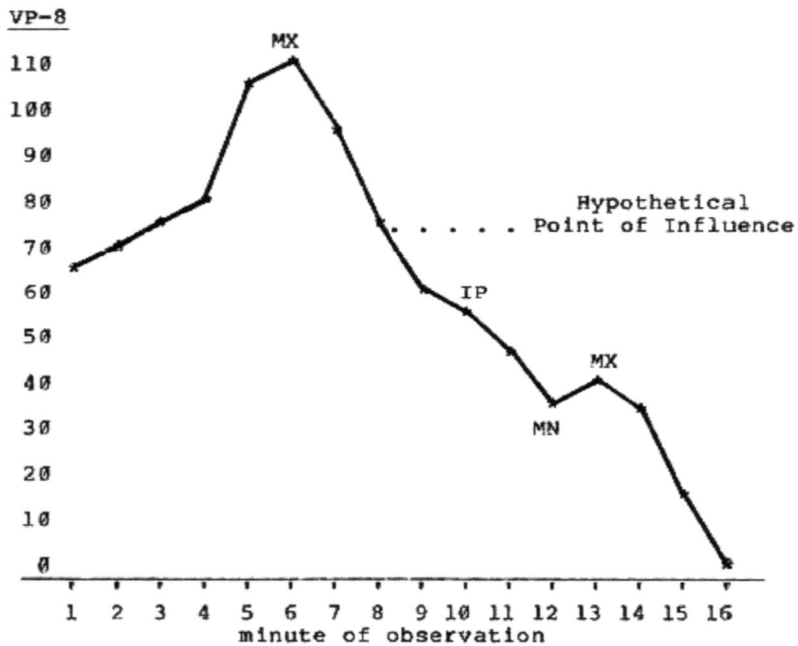

VP-8 reading:
62 69 75 79 104 110 95 77 59 55 47 38 42 37 12 0

Inked Cloud Sequences:

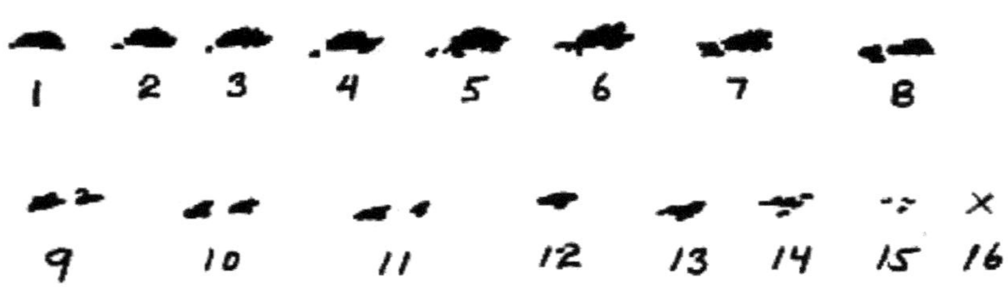

Figure 8: CONTROL SEQUENCE "U": 30 June 1978

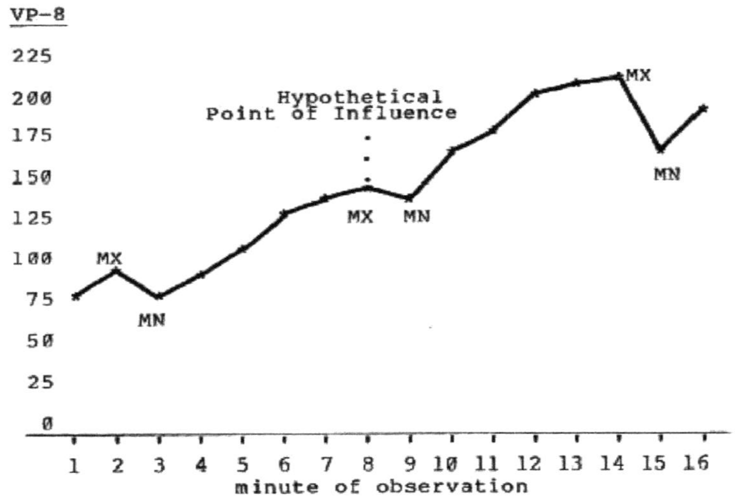

VP-8 reading:
73 91 73 86 106 124 131 146 145 165 172 197 203 207
168 181

Inked Cloud Sequences:

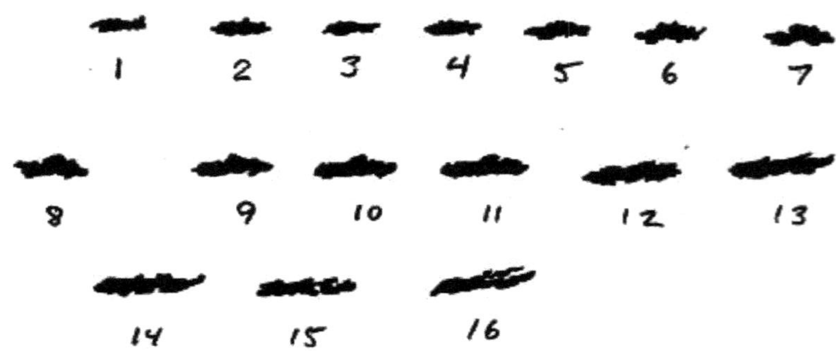

Figure 9: CONTROL SEQUENCE "V": 8 September 1978

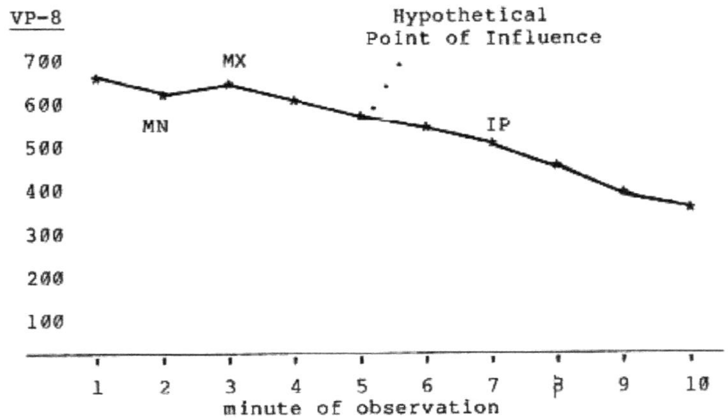

VP-8 reading:
656 625 636 618 576 535 498 467 377 349

Inked Cloud Sequences:

Figure 10: CONTROL SEQUENCE "W": 8 September 1978

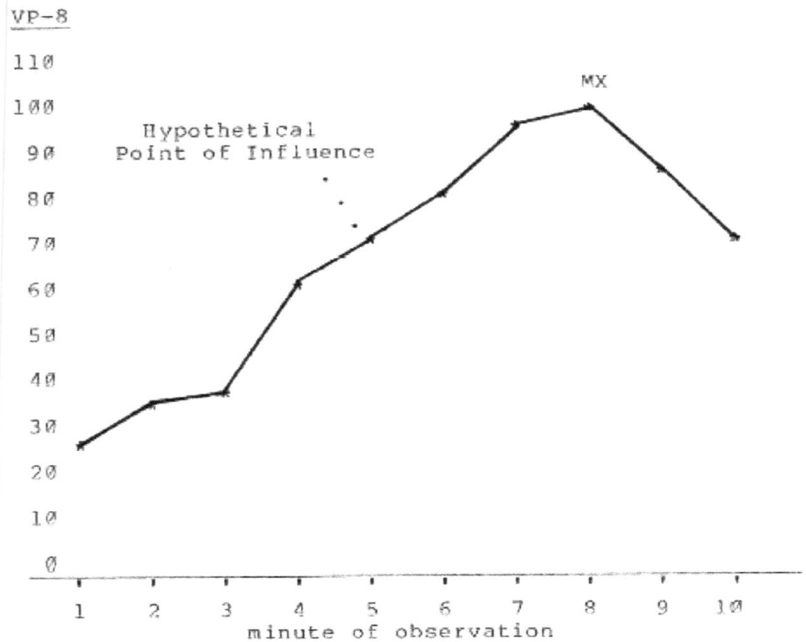

VP-8 reading:
23 33 37 58 71 79 94 97 83 70

Inked Cloud Sequences:

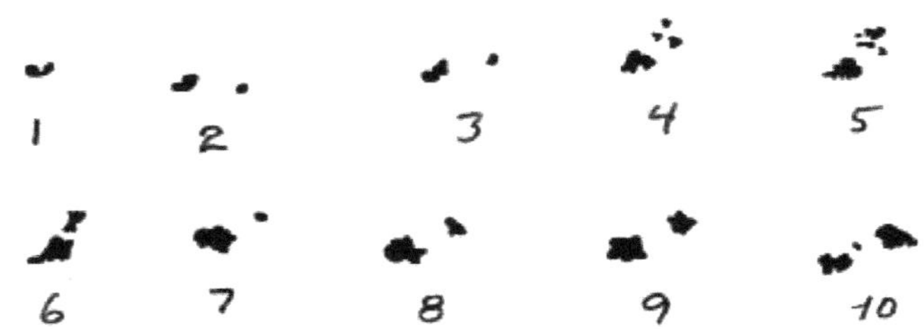

Figure 11: CONTROL SEQUENCE "X": 8 September 1978

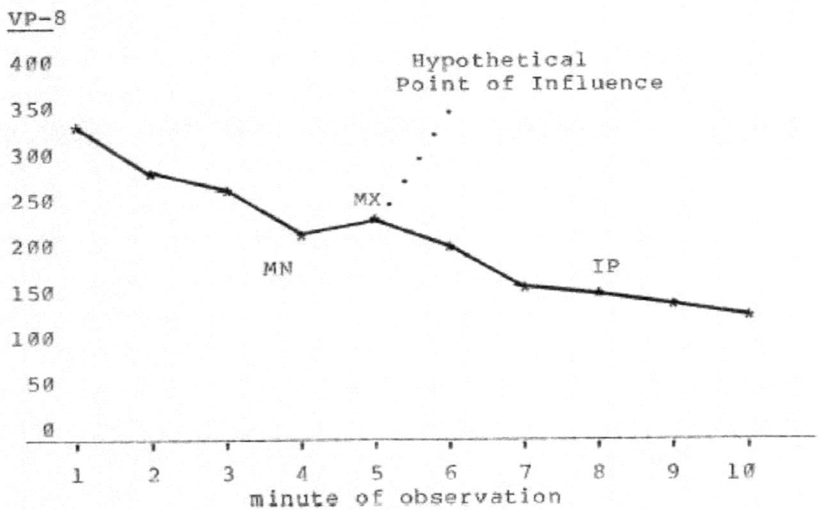

VP-8 reading:
321 275 261 215 228 195 145 140 139 123

Inked Cloud Sequences:

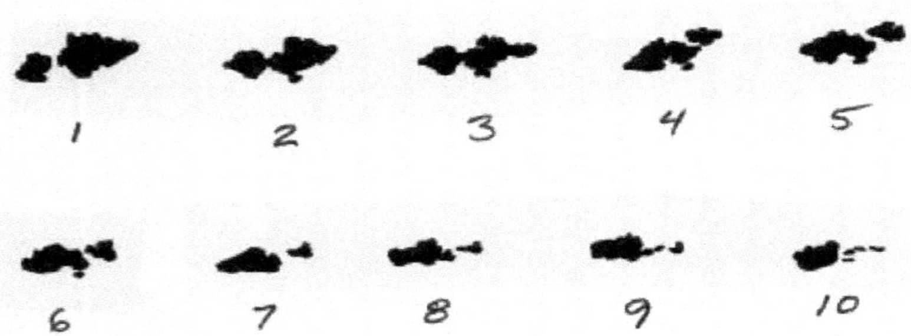

Figure 12: CONTROL SEQUENCE "Y": 8 September 1978

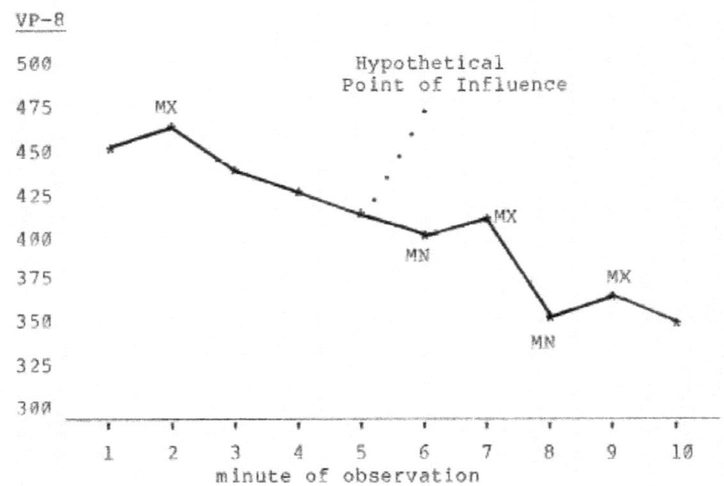

VP-8 reading:
459 465 433 424 411 404 408 357 360 398

Inked Cloud Sequences:

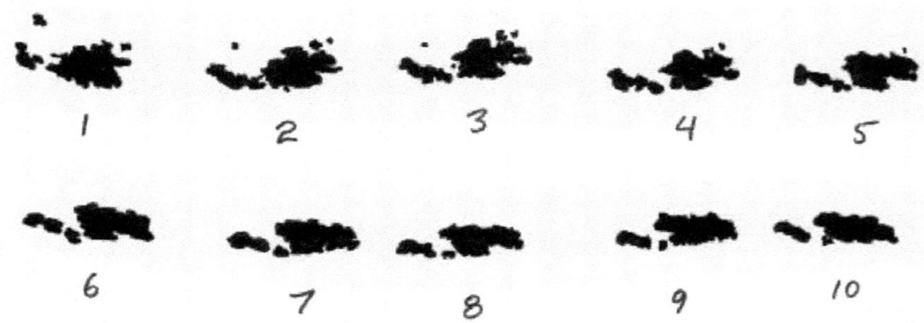

Figure 13: CONTROL SEQUENCE "Z": 8 September 1978

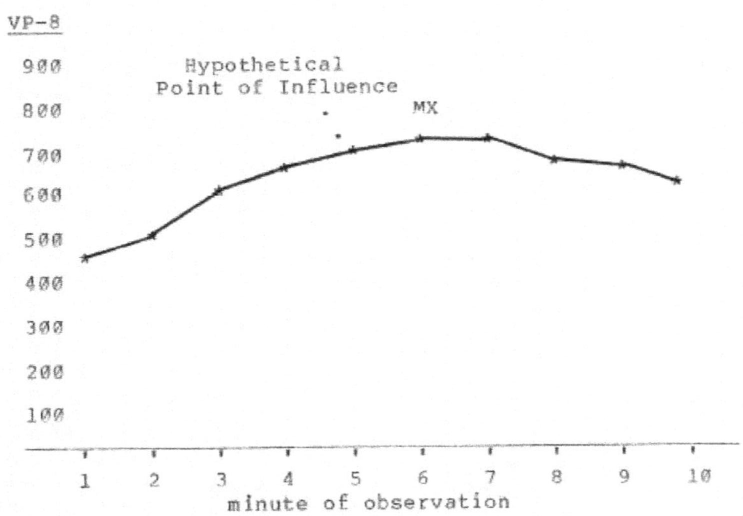

VP-8 reading:
443 511 618 638 691 711 706 677 645 633

Inked Cloud Sequences:

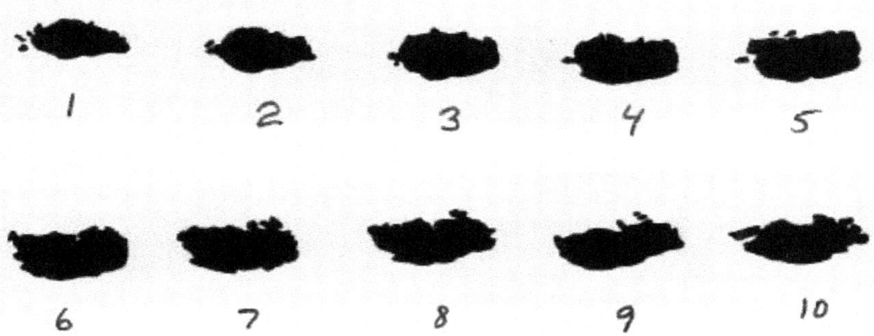

Figure 14: TESTED SEQUENCE "A": 30 June 1978

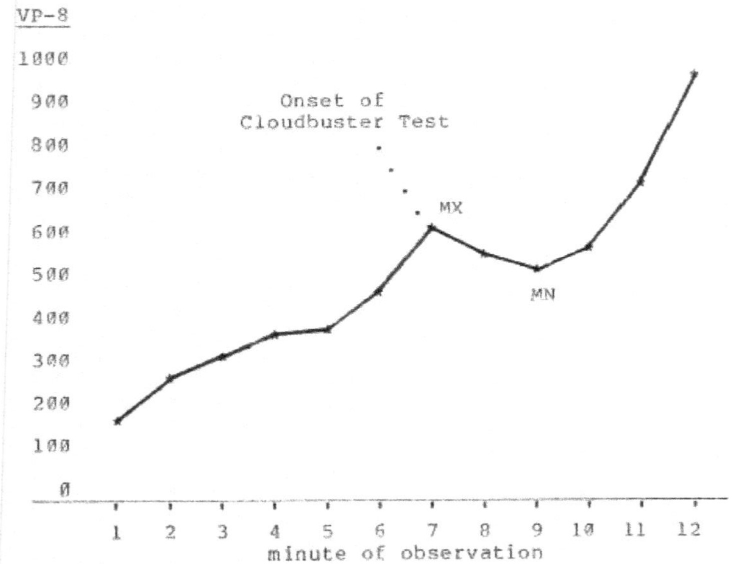

VP-8 reading:
181 245 296 240 367 451 587 529 500 542 693 920

Inked Cloud Sequences:

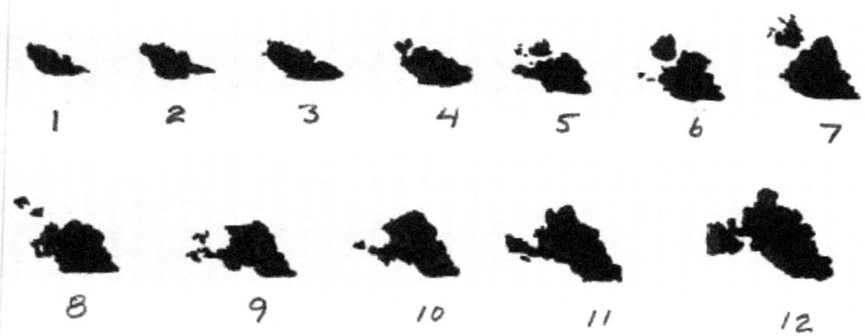

Figure 15: TESTED SEQUENCE "B": 30 June 1978

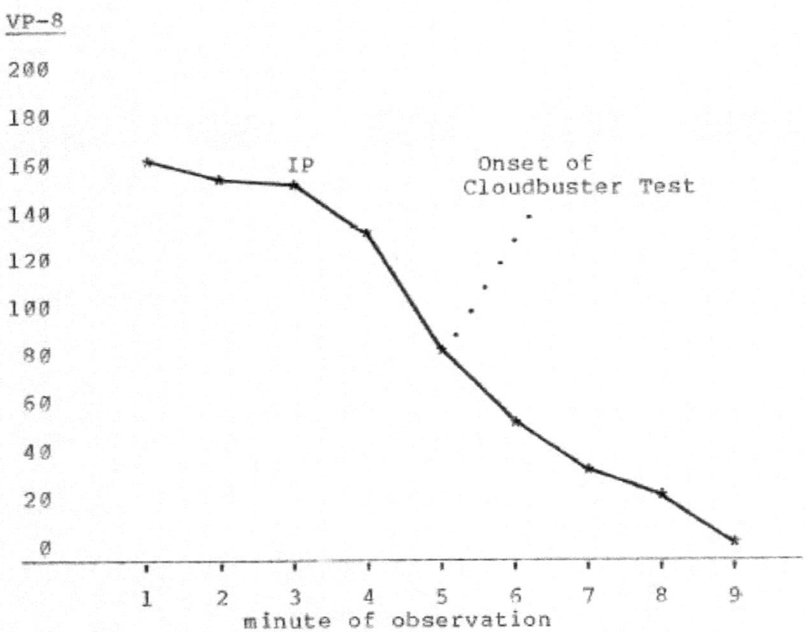

VP-8 reading:
163 157 153 134 79 53 30 19 0

Inked Cloud Sequences:

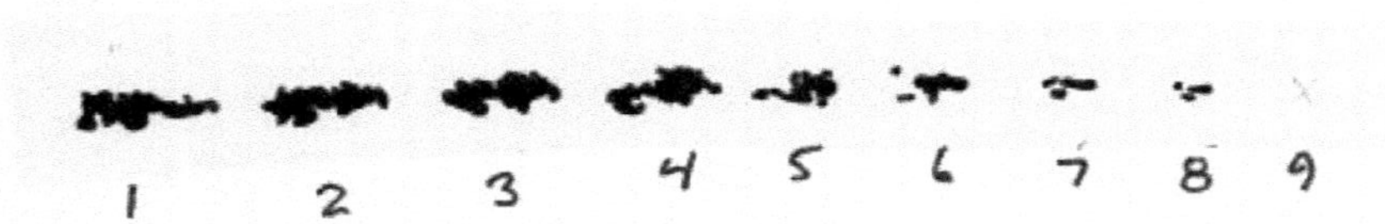

Figure 16: TESTED SEQUENCE "C": 30 June 1978

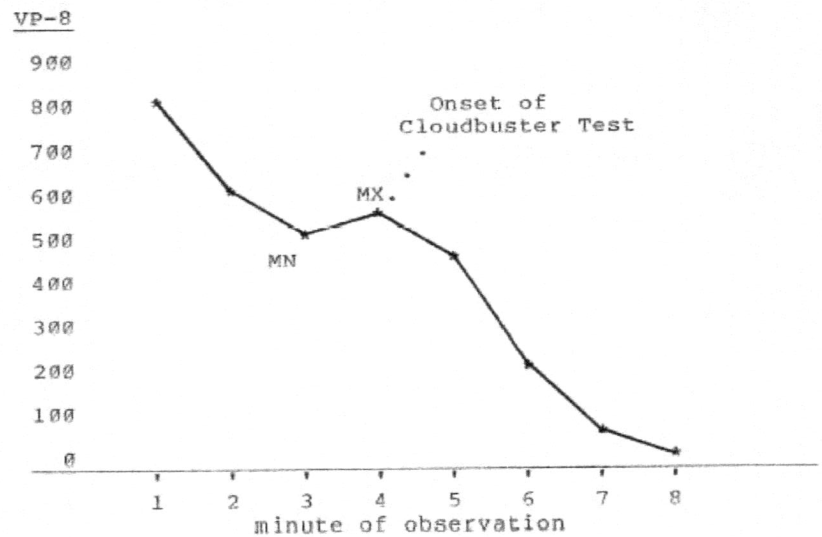

VP-8 reading:
788 610 522 561 437 171 35 0

Inked Cloud Sequences:

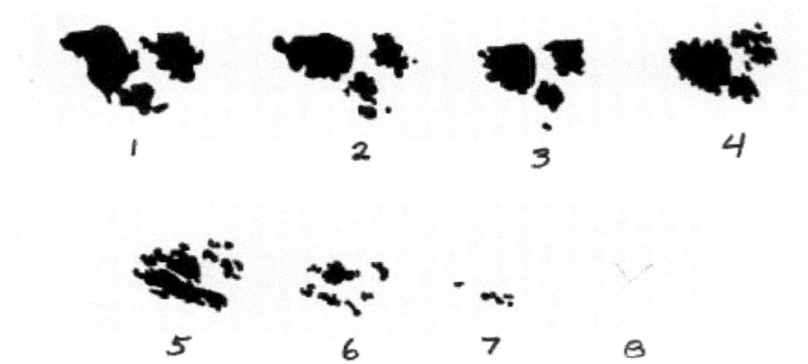

Figure 17: TESTED SEQUENCE "D": 30 June 1978

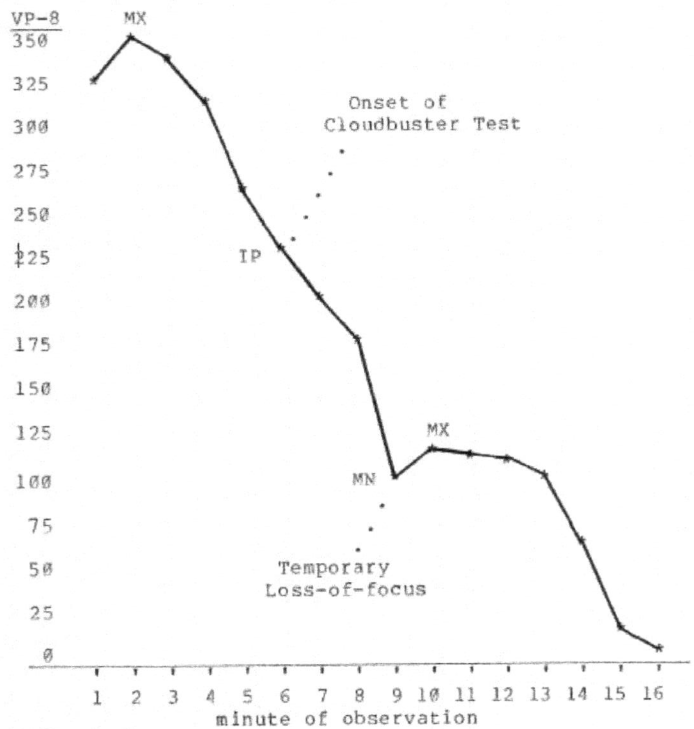

VP-8 reading:
330 347 338 321 263 230 209 171 109 116 115 113 99 70 16 0

Inked Cloud Sequences:

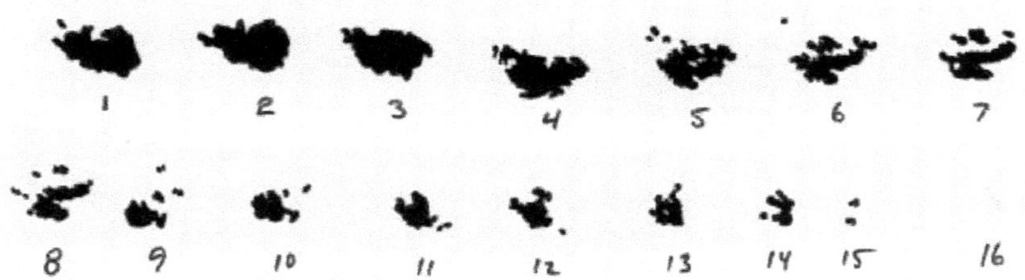

Figure 18: TESTED SEQUENCE "E": 30 June 1978

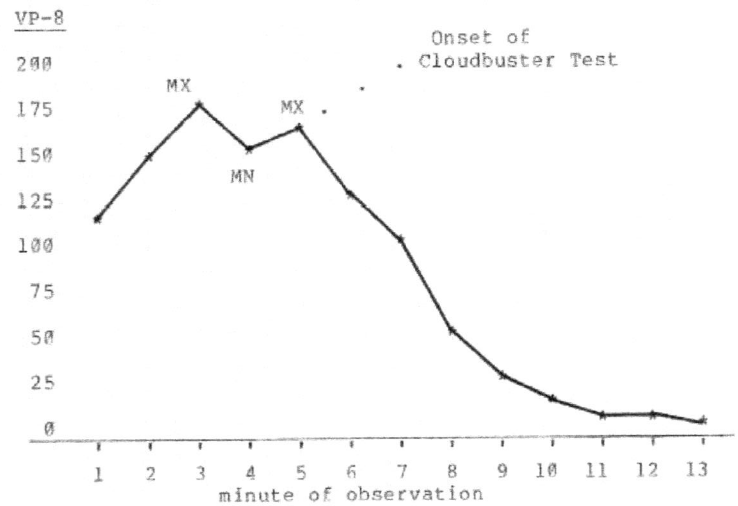

VP-8 reading:
112 146 179 152 159 125 95 56 31 17 7 2 0

Inked Cloud Sequences:

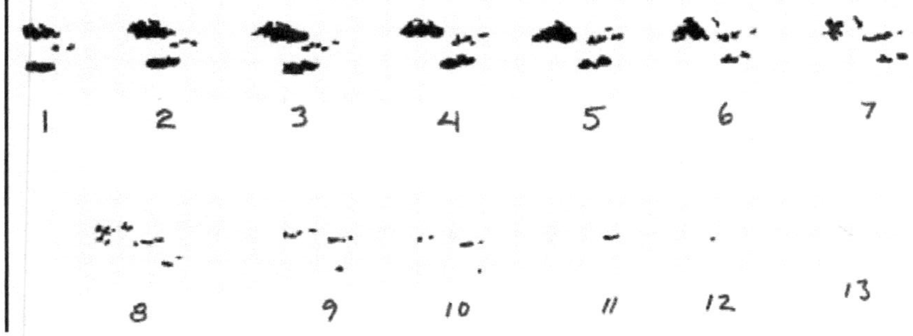

Figure 19: TESTED SEQUENCE "F": 30 June 1978

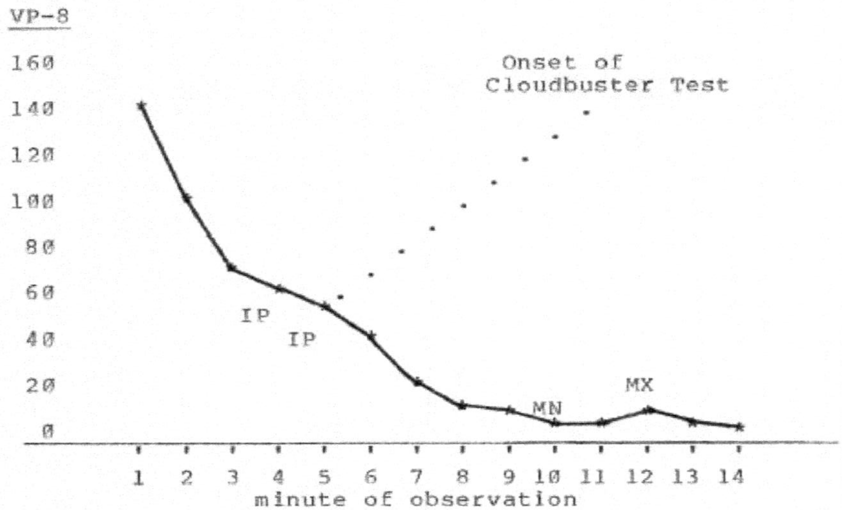

VP-8 reading:
142 108 75 60 53 40 17 10 5 4 4 5 4 0

Inked Cloud Sequences:

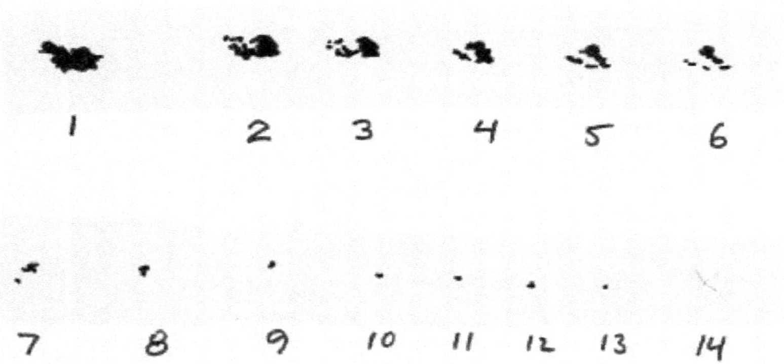

Figure 20: TESTED SEQUENCE "G": 8 September 1978

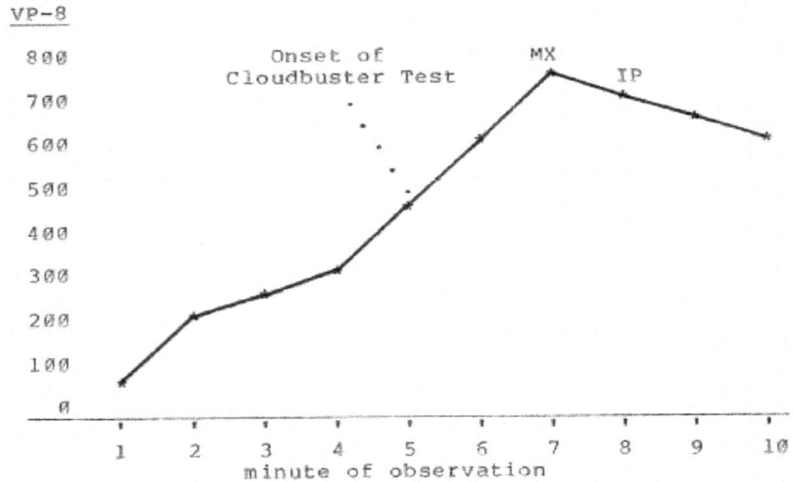

VP-8 reading:
87 187 223 300 477 600 730 711 667 636

Inked Cloud Sequences:

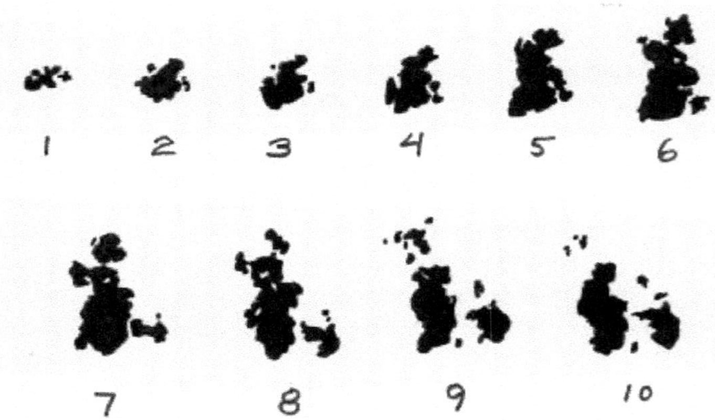

Figure 21: TESTED SEQUENCE "H": 8 September 1978

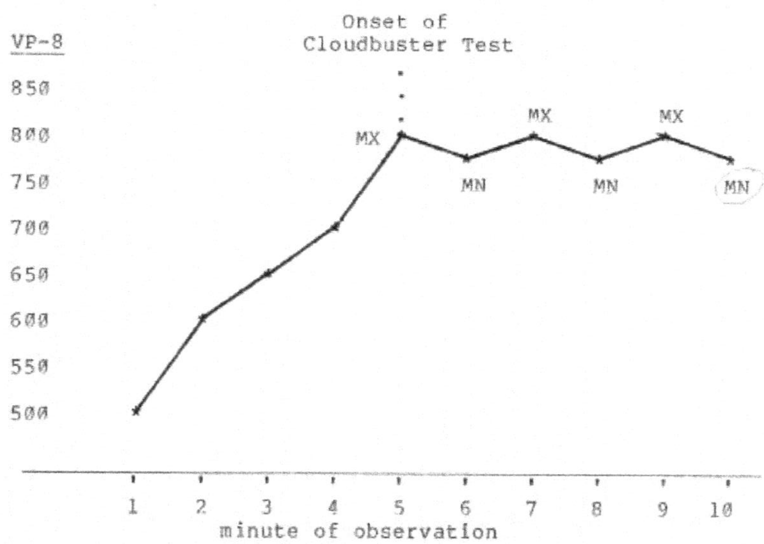

VP-8 reading:
505 589 648 709 789 773 789 772 796 752

Inked Cloud Sequences:

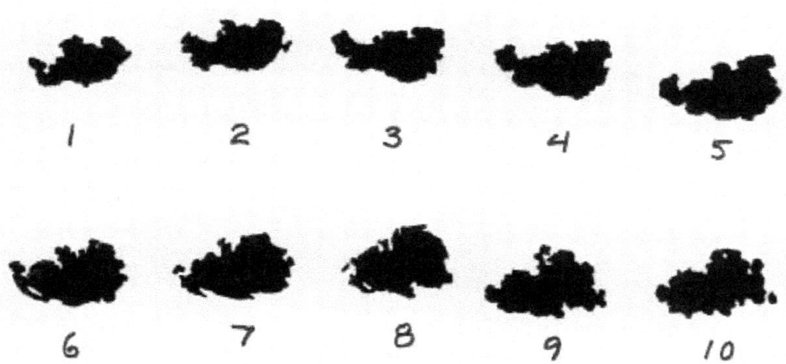

Figure 22: TESTED SEQUENCE "I": 8 September 1978

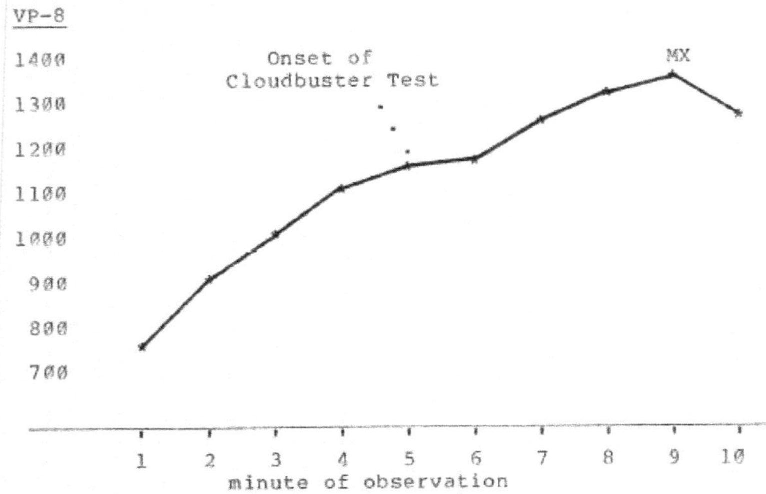

VP-8 reading:
748 889 980 1107 1161 1175 1232 1329 1337 1270

Inked Cloud Sequences:

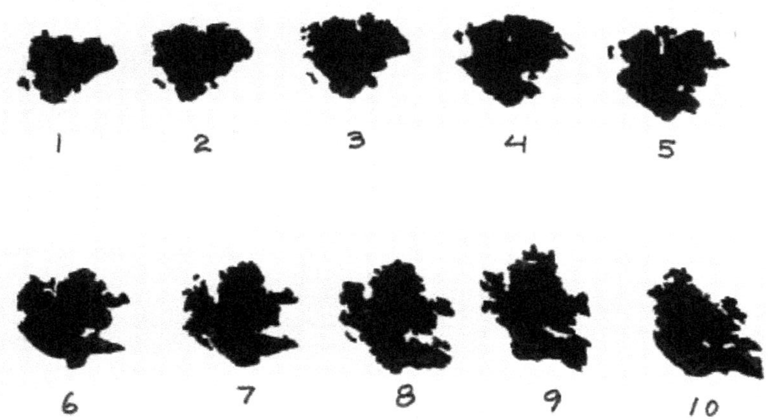

From information presented in Figures 3 - 22, we may note the following:

1. 27% of the control group had dissipated in an average of 15 minutes, starting with the time of selection and observation. However, 55% of the tested group had dissipated in an average of 12 minutes, starting with the time of selection and observation.

2. Six of the nine tested group clouds (67%) show significant changes in trend of data points at the same point in time where the cloudbuster was aimed at the cloud in question. Using the center-most data point in individual control cloud sequences, we can establish a hypothetical "point of influence" to compare with start of testing in the tested cloud sequences. Only three of the 11 control clouds (27%) show significant change in trend of data associating with the hypothetical point of influence.

If we take the ten center-most data points in each individual cloud sequence for the cloudbuster-tested group, aligned on the same point in time where the cloudbuster operation began, we can look for changes in trend of dissipation in the clouds, as a group, for the separate periods before and after the operation began. Using a trend line (best straight line) analysis, we can graphically look for changes in cloud behavior. Figure 23 graphically depicts such changes, revealing a downward trend in cloud dissipation occurring in the last half of the data. This change in trend of the data implies an increased rate of dissipation after the cloudbuster was aimed at the clouds.

In the control group, the hypothetical point of influence usefully divides the cloud sequences into "before" and "after" categories of data points which can be compared with cloudbuster tested-group data. Using the control cloud group for comparison, we find no such departure in trend for "before" and "after" a hypothetical point of influence. Figure 24 depicts this control group data, where "before" and "after" groups show nearly identical trends. Data in Figures 23 and 24 is expressed as a percent of average for each individual cloud sequence, so that each cloud would be emphasized equally. Again, we recall that control cloud sequences were photographed on the same day and immediately prior to the cloudbuster-tested sequences.

3. All data points in the control and tested group sequences (Figures 3 - 22) can be classified as follows:

A. Significant Data Points:

1) Relative Maxima (MX) — indicates a change in trend of the data from growth to decay.

2) Relative Minima (MN) — indicates a change in trend of the data from decay to growth.

Figure 23: TREND ANALYSIS OF TESTED SEQUENCES

Tested group data: 30 June 1978 and 8 September 1978

Trend before test ———————

Trend after test - - - - - - - - -

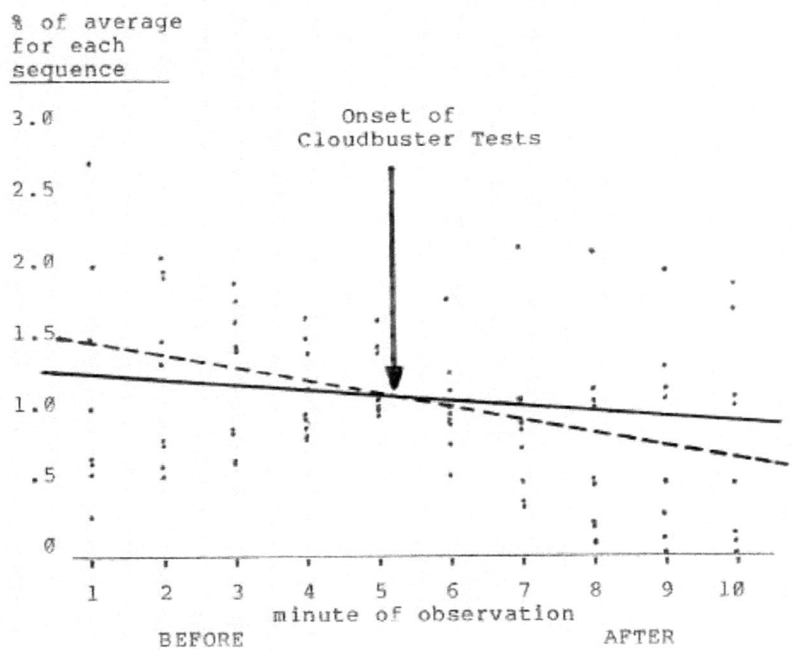

Figure 24: TREND ANALYSIS OF CONTROL SEQUENCES

Control group data: 30 June 1978 and 8 September 1978

Trend "before" ———————

Trend "after" - - - - - - - - -

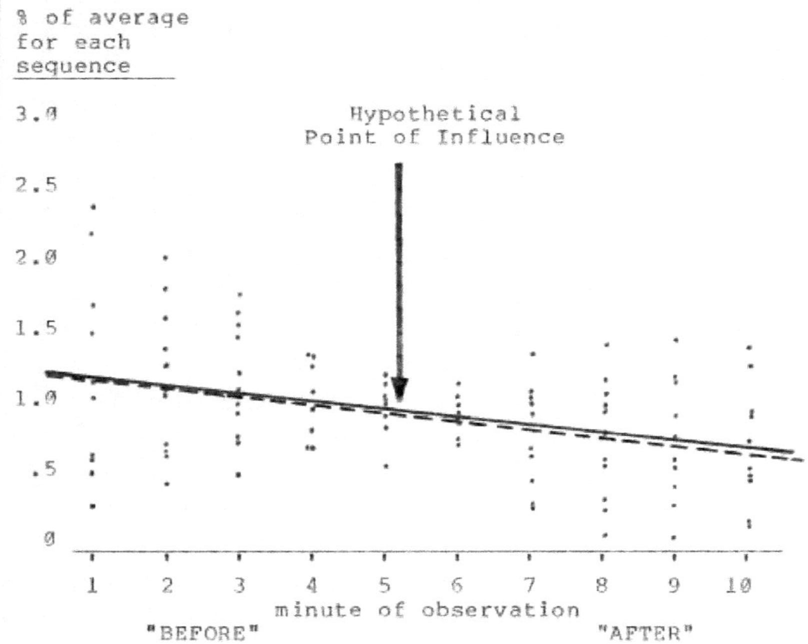

3) Downslope Inflection Points (IP) -indicates a change in trend of the data from a gradually decreasing rate of decay to an increasing rate of decay.

B. Non-Significant Data Points: none of the above changes occurring in the data.

It takes at least three data points in the graphs to define a MX or MN, and at least five data points to define an IP. These three types of significant data points are labeled on graphs in Figures 3 - 22.

Table I below summarizes the types of data points in all these graphed cloud sequences:

Table I: DATA POINT CLASSIFICATION

Type of Data Point	Tested group	Control group
Significant Data Points:		
Relative Max. MX........	12	22
Relative Min. MN........	7	15
Down. Inflec. IP........	5	9
Total MX, MN, IP........	24	46
Non-Significant Data Points:		
Total..................	78	104
Total number of data points..	102	150

Figure 25 portrays a graphic distribution of significant data points MX and IP for the 10 central- most data points in all cloud sequences, aligned with the onset of cloudbuster operations (tested group) or hypothetical point of influence (control group). One will note the largest number of significant data points occurs in the tested group sequences at minute five of the observational period.

Photographs taken at minute five, from which these tested-group data were extracted, were immediately followed by aiming of the cloudbuster at the cloud in question. Hence, minute five is the data point one would look at for changes in trend of cloud behavior, assuming an immediate response in clouds to the cloudbuster's influence. Figure 25 would indicate that more clouds began to dissipate just after minute five in the tested group sequences than at any other time, or for either group.

4. The observation of regrowth of clouds after loss-of-focus during cloudbuster dissipation tests was observed on several occasions, but only once during the photographed tests. Figure 17 shows a set of inked images for this phenomena of cloud regrowth after loss-of-focus; the event appears as a Relative Minima (MN) in the graphed data. In the tested group there were 7 MN data points out of a total of 102, indicating a low probability that this single loss-of-focus event could have chosen the MN data point by chance alone. In the tested group we do not find any MN data points associating with minute five, the onset of operations; such an association would not normally be expected in the context of our thesis, that the cloudbuster dissipates clouds. An MN data point indicates the cloud in question has stopped decaying and started to grow again. However, we do find an MN data point associating with the hypothetical point of influence in one of the control group sequences, as shown in Figure 10.

Figure 25: NUMBER OF SIGNIFICANT DATA POINTS PER MINUTE OF OBSERVATION

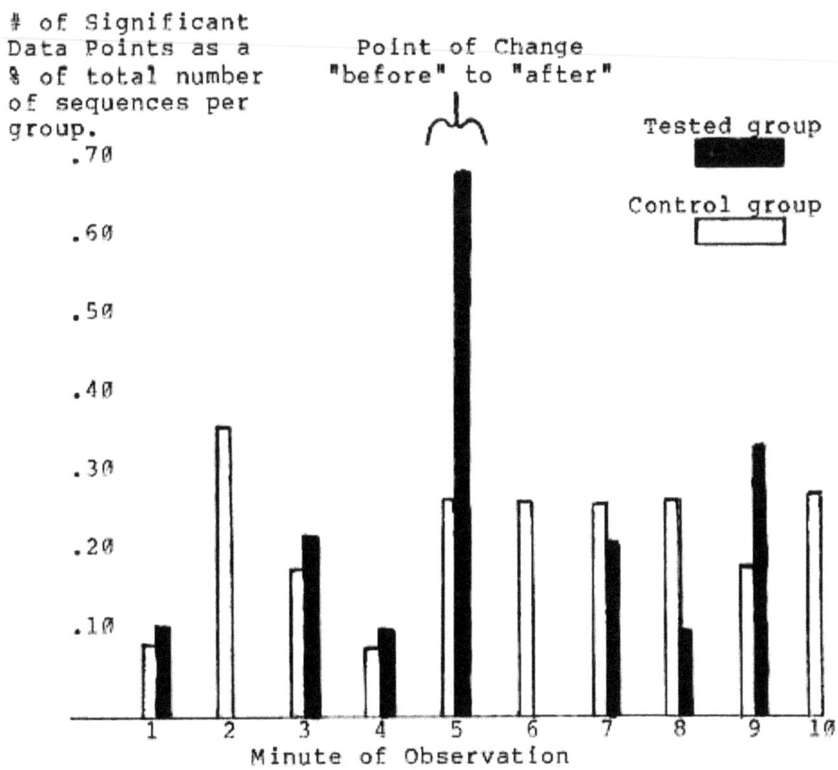

5. Using data summarized in Table I, we can compute the probabilities for the events in control and tested groups to have occurred as they did. We use a calculation of probability for partitioned events (Lentner 1972), where:

$$\text{probability} = \frac{\binom{N_1}{n_1}\binom{N_2}{n_2}\cdots\binom{N_k}{n_k}}{\binom{N}{n}}$$

$$\text{and where } \binom{N}{n} = \frac{N!}{n!\,(N-n)!}$$

In the above equation:

N = total # of data points
n = number of sequential observations
N_1 = total number of RM in data
n_1 = total number of RM at minute five
N_2 = total number of IP in data
n_2 = total number of IP at minute five
N_3 = total number of non-significant data points
n_3 = total number of non-significant data points at minute five

therefore,

$$\text{probability (tested group)} = \frac{\binom{12}{4}\binom{5}{2}\binom{78}{3}}{\binom{102}{9}} = 0.00016$$

and,

$$\text{probability (control group)} = \frac{\binom{22}{2}\binom{9}{1}\binom{104}{8}}{\binom{150}{11}} = 0.05$$

The probability of the control test sequence occurring is p = 0.05, or about one chance in 20, largely due to "hypothetical selection" of a large number of non-significant data points. However, the probability for the tested group sequence to occur is p = 0.00016, or about one chance in 6000. Occurrence of the tested sequence of data is about 300 times less likely than the control sequence of data. One does not normally expect a cloud to dissipate by chance alone at a pre-selected moment for any significant number of consecutive times. In the trials here, cloud volume began to decrease (or increasingly decrease) in six of nine trials (67%) at the same point in time that the cloudbuster was aimed. The control group showed "hypothetical selection" of significant data points MX and IP in only three out of 11 trials (27%). It should be noted that the control group had a greater population of significant data points to "select" from (21% of all data points) than did the tested group (17% of all data points).

Chapter III

TESTS FOR PRECIPITATION ABATEMENT

During the test period of 1977 - 1978, three separate attempts were made to reduce the intensity of local storms. On two of these dates, 25 June 1977 and 16 April 1978, weather was exceedingly rainy for the previous days. In each case a decision was made to attempt a wide scale dissipation of the local storm systems after flash flood warnings went out; operations commenced in the afternoon of the dates above. In these cases the cloudbuster was scanned over a wide area where the center of the storm system was roughly determined to be. The hope was to trigger a dissipation similar to that noticed in the cumulus cloud experiments, only on a larger scale. The third attempt, made on 23 May 1978, was targeted to reduce the intensity of a locally severe thunderstorm which was heading toward Lawrence, Kansas from the S.W. This operation also commenced late in the afternoon, after a tornado was reported in the storm.

Appendix V contains a duplication of log notes for some of the operations conducted. These notes, in conjunction with materials in Appendix III, will hopefully convey a sense of how the cloudbuster is operated during a major operation, and the kinds of changes in the atmosphere one looks for during them. In reading these appendices, one should keep in mind that instruments for measuring and recording the reported phenomena were not available to this researcher and, for some of the phenomena, has yet to be adequately developed.

General Observations

While the total number of precipitation abatement operations is small, several general observations were noted that had been previously reported by other researchers.

1. Increased brightness at ground level shortly after commencement of operations — a "sparkle" returns to the air. I sensed an increase in "brightness discrimination" to occur, much as reported by researchers with negative air ions and as subsequently experienced by myself with a small negative ion generator. In these cases, however, there also occurred a simultaneous slight breakup in overcast and haze. The borders of clouds in a previously homogeneous overcast become discernible — the clouds began to aggregate, sometimes enough to let sunshine through.

2. Abatement of wind intensity with operational focus toward specific compass directions.

3. Triggering of wind at onset of operations when no wind was present beforehand. Such winds were particularly striking when the atmosphere had been "still" for several hours prior to the operation. Gentle air movements and sometimes gusts seemed to spring up at onset of the operation.

4. Biological reactions of birds, which would be rousted from silence and often begin flying around the device.

5. Negative biological reactions of the operator, from accidental skin contact with the cloudbuster during its operation. These include extreme fatigue and muscular cramps of sustained duration, plus mild to severe nausea, diarrhea and hives. Such ailments are normally foreign to the character of the operator and would be felt to "hit" the body a moment after the careless act. The cloudbuster is generally operated with gloves and on a remote control cable to avoid such reactions, which were noted by all who work with the device.

6. On a few overcast, non-rainy "bleak" days, when testing the device for its general cloud dissipation affect, observations 1, 3, and 4 above would occur simultaneously.

These effects were generally not documentable, nor apparently widespread enough to manifest in meteorological data from National Weather Service offices in nearby Kansas City or Topeka (about 50 miles distant). These observations are admittedly subjective in character, but it is precisely these subjective phenomena that the cloudbuster operator initially looks for in terms of operational effect. It must be remembered that many "objective" phenomena were initially someone's "subjective" perception, eventually made manifest on an instrument.

Data Analysis

Kansas weather data for the three precipitation abatement operations combined is summarized in Figure 26; the data for the operations individually is found on Figures 27, 28, and 29. These data all indicate a decrease in daily precipitation values over the State for at least two days following the cloudbuster operation. Time of day analysis does not indicate a clear association between onset of operation and hourly precipitation or % cloud cover. There is, generally, less precipitation occurring in the 40 hours after the operations than in the 40 hours preceding them. Also, we note a relative absence of diurnal cycling in % cloud cover for the three days prior to the cloudbusting operations. However, shortly after the cloudbusting operations began we notice a general restoration of diurnal pendulation in % cloud cover (see Figure 26).

Figure 26: WEEKLY SUMMARY -- THREE PRECIPITATION ABATEMENT OPERATIONS

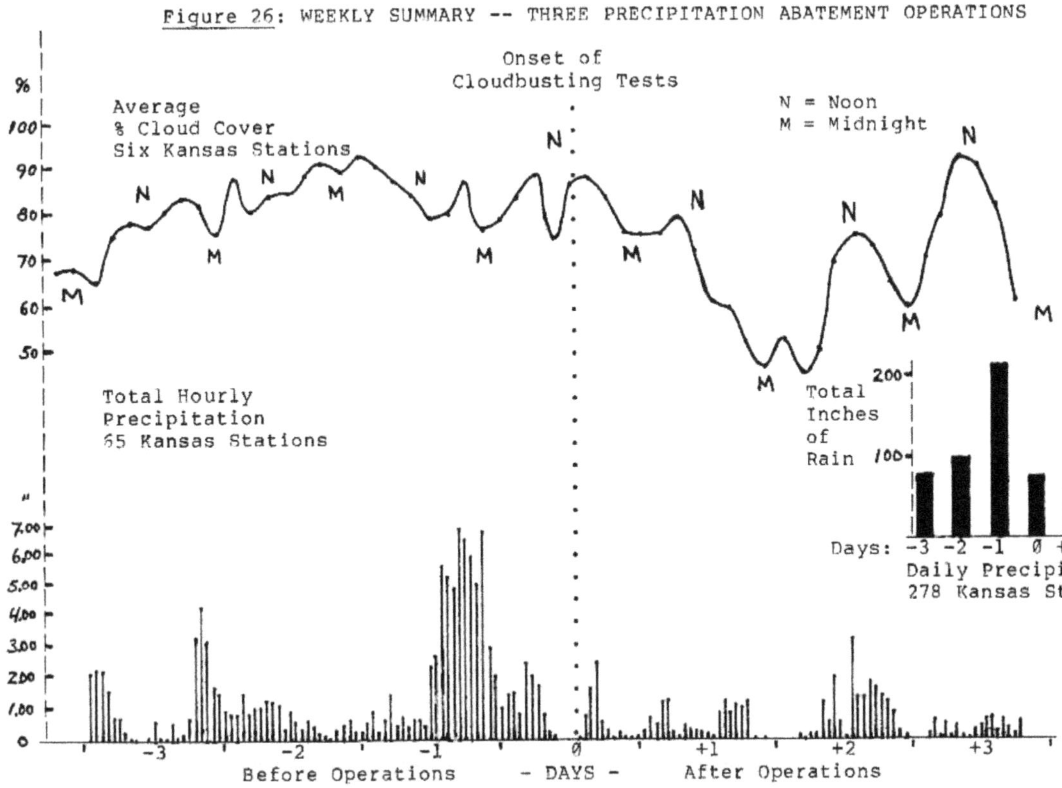

Figure 27: PRECIPITATION ABATEMENT OPERATION -- 25 June 1977

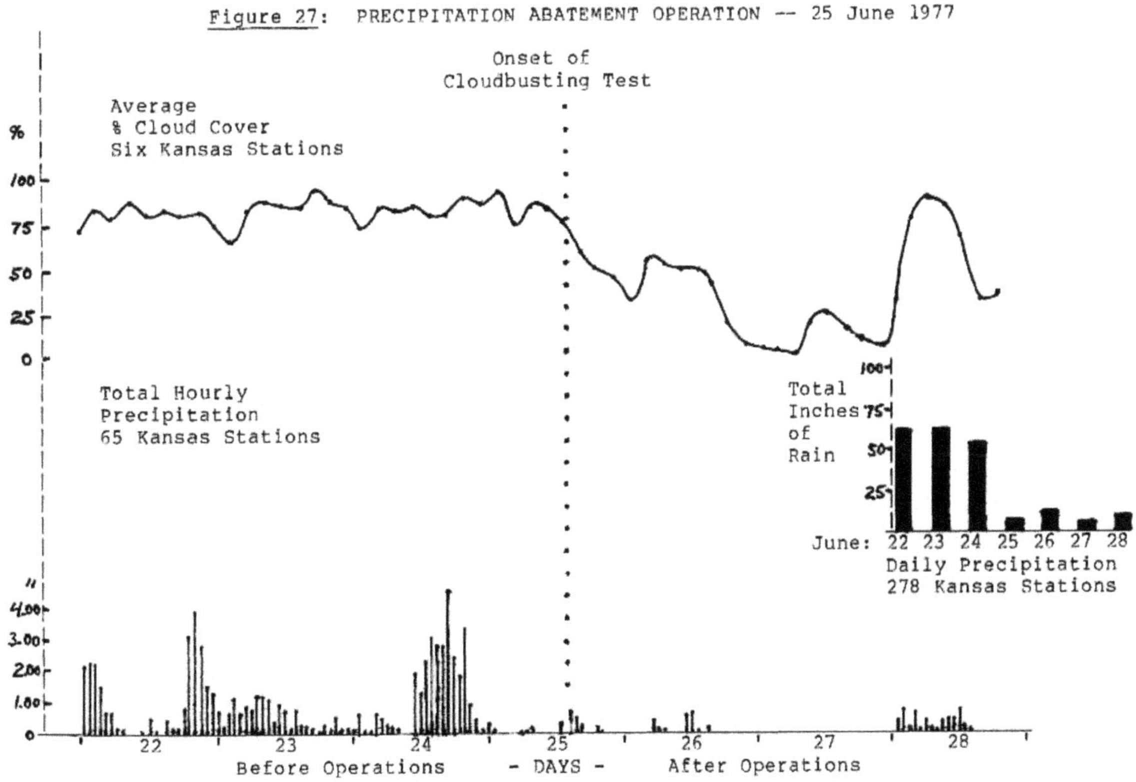

Figure 28: PRECIPITATION ABATEMENT OPERATION -- 16 April 1978

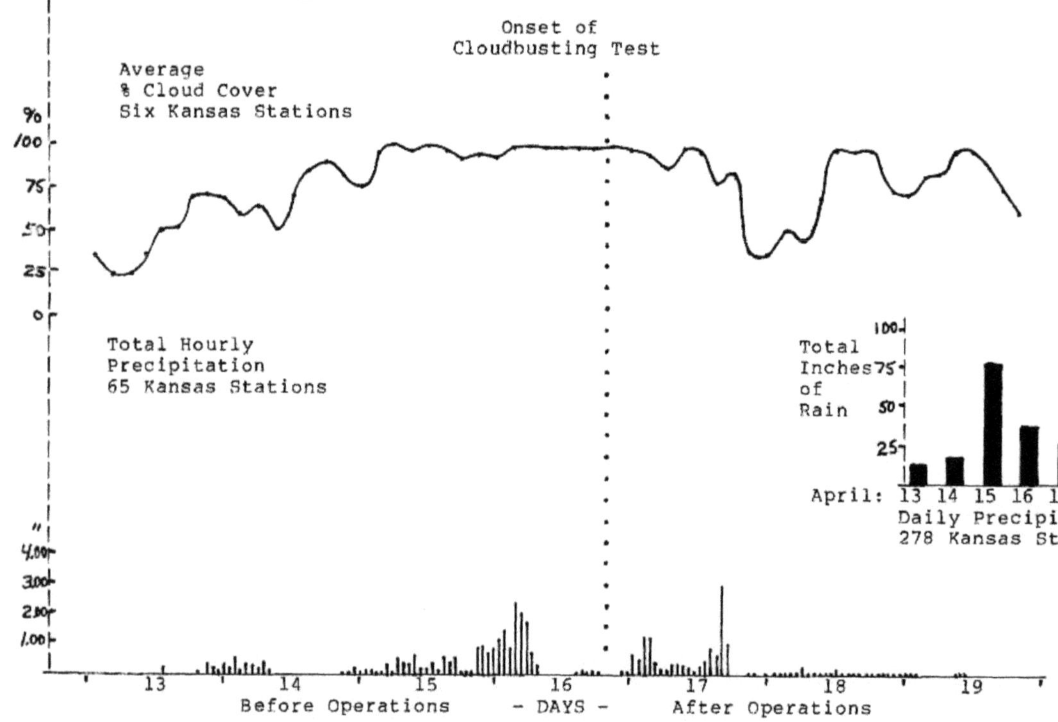

Figure 29: PRECIPITATION ABATEMENT OPERATION -- 23 May 1978

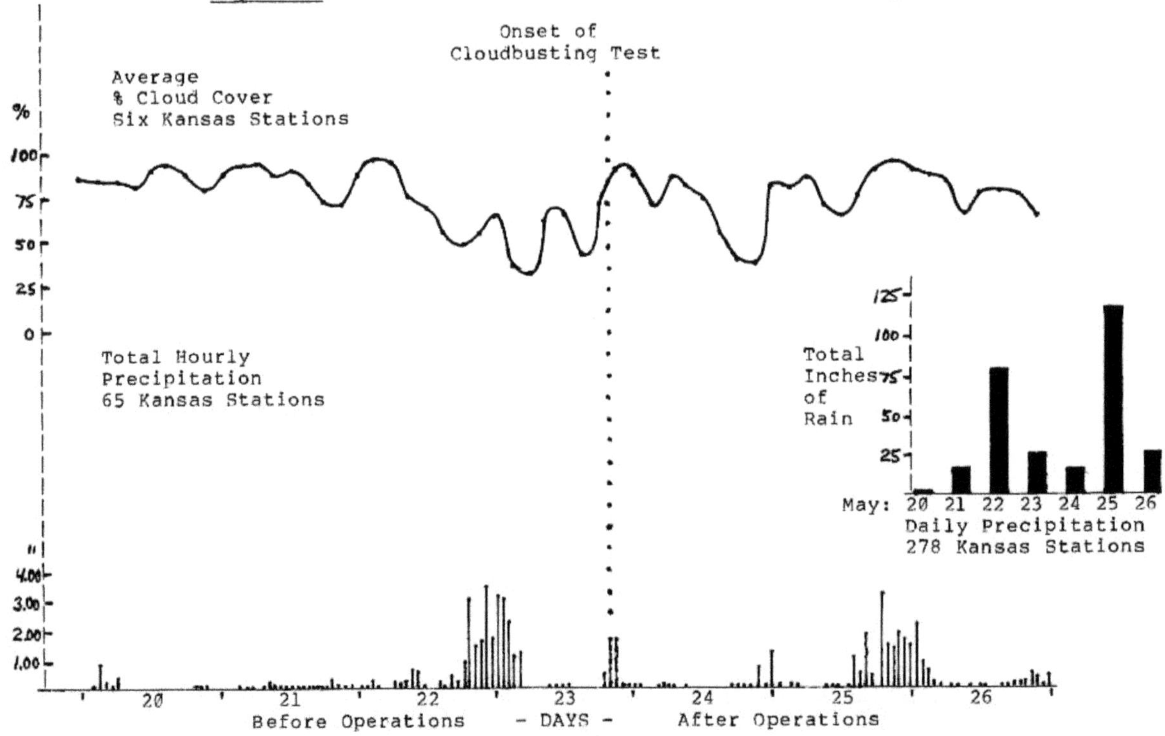

It is difficult to make other than subjective observations from these data. For instance, it was learned retrospectively that the National Weather Service had predicted a clearing of weather on the day after the operation of 25 June 1978 and, just such a clearing occurred. A clearing was not predicted by N.W.S. for the other two operations and, as indicated in Figures 27, 28, and 29, the clearing which did occur on these two dates was much less marked. Satellite photographs of the central U.S.A. did not show any clear changes associated with time of the cloudbusting operations.

Regarding the cloudbusting operation of 23 May 1978, in which I attempted to de-intensify a severe storm with a tornado in it, the following observations were made. At 6:30 pm on this date Lawrence was under a severe storm watch and the sky was cast with an eerie yellow-red. The winds were extremely calm ahead of the storm, and local birds alternated from chaotic noisy chatter to absolute silence. The operation started at 8:45 pm whereupon the storm seemed to increase its lightning output at high altitudes. It also seemed to loose some of its distinctive thunderhead "anvil" organization and veer northward, missing Lawrence and Kansas City. An increase in lightning, for whatever reasons, would leave the storm with much less energy to expend in other directions, such as in maintenance of a vortex. Storm Data indicates that this severe storm generated a tornado which touched down in Osage County, S.W. of Lawrence, with a path of about 4 miles. No indication is given in this data as to when the storm de-intensified, however, since the tornado was short-lived and began at around 6:30 pm, it seems certain that the storm had already changed in character by the time my operations had begun at 8:30 pm. No severe weather was reported in the area for the remainder of the evening well into the next day. Daily weather maps for the morning (7:00 am E.S.T.) of the day of operations, and for the following morning, are included in Appendix IV. No exceptional changes in patterns of frontal movements or in the elevation of the 500 mb pressure surface was noted for these weather maps, other than slight increases in southerly air flow manifest as a warm frontal movement northward.

CHAPTER IV

TESTS FOR PRECIPITATION ENHANCEMENT

We now turn our attention to a section that relates the result of 12 separate operations designed to induce or enhance precipitation. The study area is, again, the state of Kansas. Dates and starting times of the 12 operations were:

	date	time (C.S.T)
1.	14 July 1977	3:00 pm
2.	17 July 1977	2:00 pm
3.	30 July 1977	2:00 pm
4.	8 September 1977	5:30 pm
5.	26 August 1978	12:15 pm
6.	3 September 1978	12:55 pm
7.	10 September 1978	3:00 pm
8.	17 September 1978	12:20 pm
9.	24 September 1978	12:40 pm
10.	1 October 1978	12:25 pm
11.	8 October 1978	12:35 pm
12.	15 October 1978	12:35 pm

With the exception of 17 July 1977 and 15 October 1978, the above 12 operations were performed using a combination of the following techniques:

1. "Zenith draw" — the cloudbuster tubes are aimed directly overhead. This is said to contract downward the earth's orgone energy envelope, inducing cloud formation and precipitation. The technique was employed from one to five hours per operation.

2. "Draws" to various compass directions — usually to the N. W. horizon up to 30' elevation. This is said to induce frontal movement and intensification from the direction of the draw, usually within 24 hours of the operation.

Operations on the two dates previously noted used a zenith draw only; this seemed to have had a noticeable difference in effect, as we shall see.

A fundamental point is that one does not go out with a cloudbuster to "make rain". Cloudbusting literature is clear on this point. The goal is to remove obstacles in the atmosphere which prevent natural cycles of rainfall from occurring. Some of these 12 operations, it may be noticed, broke with strict observance of the above consideration; the operator takes full responsibility for any risks involved. All eight 1978 precipitation enhancement operations were planned in advance, to occur a week apart on successive Sundays. The goal was to induce an artificial periodicity in precipitation.[1]

These pre-planned operations were performed with the stipulation that if any possible risk of triggering violent weather existed, the tests for that day would be cancelled, irrespective of what it would mean for the outcome of the periodicity experiment.[2] As we shall see later, none of these tests required cancellation, nor were any increases in severe weather phenomena associated directly with the operations. A Bearcat Weather Radio, capable of receiving National Weather Service emergency broadcasts in Kansas was monitored during most of these tests.

General Observations

1. On cloudless days, after rapid cloudbuster draws were made in opposite compass directions of the sky, small patches of cirrus billows or altocirrus were often found to appear in the nearby sky. This was observed on several occasions. However, these clouds would shortly begin to drift out of the area, and I was never able to see one of them develop to any further extent. Zenith draw did seem to preserve them.

Photo Group 5 shows four of these clouds. It should be noted that a moment before the test started these clouds were not present, and a moment after the operation started they were forming and developing, never to grow larger than as seen in the photographs. The photos may make some of these clouds seem deceptively widespread, however, the photos were taken at each clouds closest passage. These clouds formed generally upwind of the cloudbuster site, and opposite to the direction of the draw.

2. On many of the operational days, the sky was clear blue at the onset, without a cloud to be seen. Shortly after initiation of the zenith draw, a haziness would set into the atmosphere. This haziness would slowly aggregate and eventually blossom into small, widely-dispersed cumulus clouds. These clouds would grow and decay, but generally increase in number and size with passing time

---

1. Similar tests, using cloudseeding techniques, were first reported by Langmuir (1962), the "father" of cloudseeding. Langmuir seeded the atmosphere over New Mexico once a week and noted an atmospheric response which extended throughout "the whole of the Northern Hemisphere" (op.cit. Vol 11, p.xx). Langmuir's work was rejected by his contemporaries on statistical grounds (Brier 1955), but his experiments have yet to be repeated.

2. We do not wish to add cloudbusting to an already long list of innovations used in a socially irresponsible fashion under the guise of "science for science's sake".

eventually to scatter widely across the sky. Sometimes this is all that would develop within eyesight of the operator. At other times, cumulus would further organize into streets. These cumulus streets would then either grow and develop into towering cumulonimbus,[3] or temporarily decay away leaving the sky clear, only to display, shortly thereafter, towering thunderhead tops on the horizon. These thunderheads were usually W. or S.W. of the cloudbuster site, as were small cirrus formed via other experimental methods mentioned above. Often a high layer of altostratus would grow at the same time cumulus streets were forming. Oscillation between altostratus and cumulus would occur, with the pendulation gaining in intensity as time passed. This pendulation would also result in thunderheads on the horizon. Photo Sequence 6 shows this full transition, from cloud-free sky through development of altostratus and cumulus streets, to thunderheads in the early evening.

3. On occasion, the operator experienced a "pressure" or "tension" sensation inside the head during zenith draws. These sensations were, on 3 September 1978, associated with a measured rise in barometric pressure and a simultaneous increase in % cloud cover (cumulus count) at the operational site. These data are contained in Figure 30. At the start of operations on 3 September 1978 there were no clouds anywhere in the sky. I can think of no classical explanation as to why such large changes in barometric pressure would associate with the observable number of cumulus clouds, as well as pressure sensations in the head. This is particularly so when one considers a pressure rise was associated with increased cumulus, both variables increasing immediately after start of operations. The cumulus count lagged behind the pressure rise, which dropped off the moment the operation was terminated. Waxing and waning of head pressure sensations also associated with commencement and termination of operations. A secondary peak and dip in the data associates with a minor restart and stop of the operation later on in the day. The operation terminated with large anvil tops to the S.W. of the operational site, a pattern very similar to that seen in Photo Sequence 6. This measurement technique, which proved most difficult to run single-handedly, was not repeated for subsequent operations, as it was expected that the phenomena would show up in weather records for nearby stations. Unfortunately, this was not the case, the effect seemingly confined to near the test site. However, as will be related shortly, an interesting drop in height of the 500 mb pressure surface was noted for almost all of the operations. Data in Figure 30 was gathered by the operator using a Swift barometer (model #477). Cumulus counts were made visually, for the entire 360' area around the cloudbuster; larger numbers are therefore approximate.

---

3. The role of cumulus streets as "feeders" for thunderstorms has been previously explored by Bates (1962).

4. Biological responses in the cloudbuster area were similar to those noted during precipitation abatement operations. Birds would often become excited during the tests. However, a more frequent observation was the increase in number of high-flying birds, such as hawks or buzzards, after commencement of zenith draw. A "bird count" was never performed, but this general observation would be understandable if the cloudbuster was somehow inducing instability into the atmosphere. These birds would not be flapping their wings, but merely riding the thermals, sometimes at incredible heights!

Figure 30: BAROMETRIC PRESSURE, CUMULUS COUNT AND ZENITH DRAW ASSOCIATIONS

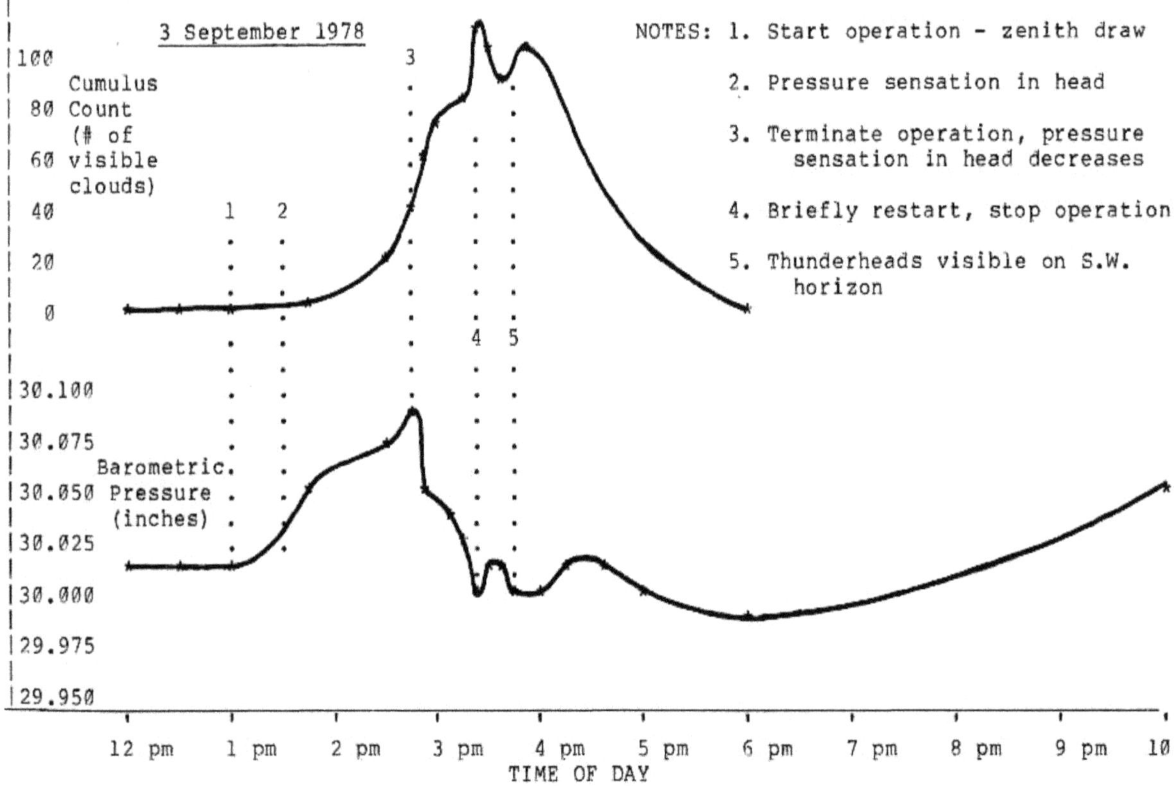

Photo Group 5: CIRRUS AND CIRROSTRATUS FORMATION

All of the above four photographs are oriented toward the W. or S.W. of the operational site, which is where the clouds initially formed. The cirrostratus formation in photo #4 does not cover the whole sky, as it may appear, although it does cover the entire photo. These formations are the only ones that could be seen in the sky at the time, and formed within a minute after commencement of the tests.

1 2 3

4 Photo Sequence 6: Precipitation Enhancement 5 6

Photo Sequence 6:
ZENITH DRAW — PRECIPITATION ENHANCEMENT

These photos show the cloudbuster set up in zenith draw position. The sky is cloud-free in photo #1 and progresses to slowly build up cloud cover. By photo #5, this cloud cover has organized into cloud streets; shortly thereafter, cumulonimbus tops advance over the horizon from the southwest, as shown in photo #6. These photos are typical of observations during a number of the precipitation enhancement tests. Photos depict changes which occur about one-half to one hour apart.

Satellite Imagery

Satellite images of the Great Plains made during these operations revealed the following phenomena:

1. In four (33%) of the twelve operations, there was noted a blossoming of isolated or frontal convective cells within one-half hour after commencement of cloudbuster operations. These occurred within 100 to 500 km of the operational site, as shown in satellite Photo Sequences 7, 8, 9, and 10.

2. In two (17%) of the twelve operations a similar blossoming of convective activity was noted within a few hours after operational commencement, as shown in satellite Photo Sequences 11 and 12.

3. In five (42%) of the twelve operations there occurred movement of intensifying frontal systems into the area within 24 hours of operational commencement, as seen in satellite Photo Sequences 13, 14, 15, 16, and 17. In these cases the frontal system often seemed to blossom only after having gotten within 100 - 200 km of the operational site.

4. No satellite photo activity was seen to associate with the operation on 15 October 1978. At that time the U.S.A. was cloud-free, from the pacific ocean to the Mississippi River, as shown in Satellite Photo 18.

Photo Sequence 7: (on following page)
SATELLITE IMAGERY — 14 July 1977

In this sequence of six satellite photos, operations started at 3:00 pm, about one half hour before photo #3 was taken. In photo #3, we note the blossoming of convective cell activity (see arrow) which, by 7:31 pm, has covered the N.W . portion of Kansas.

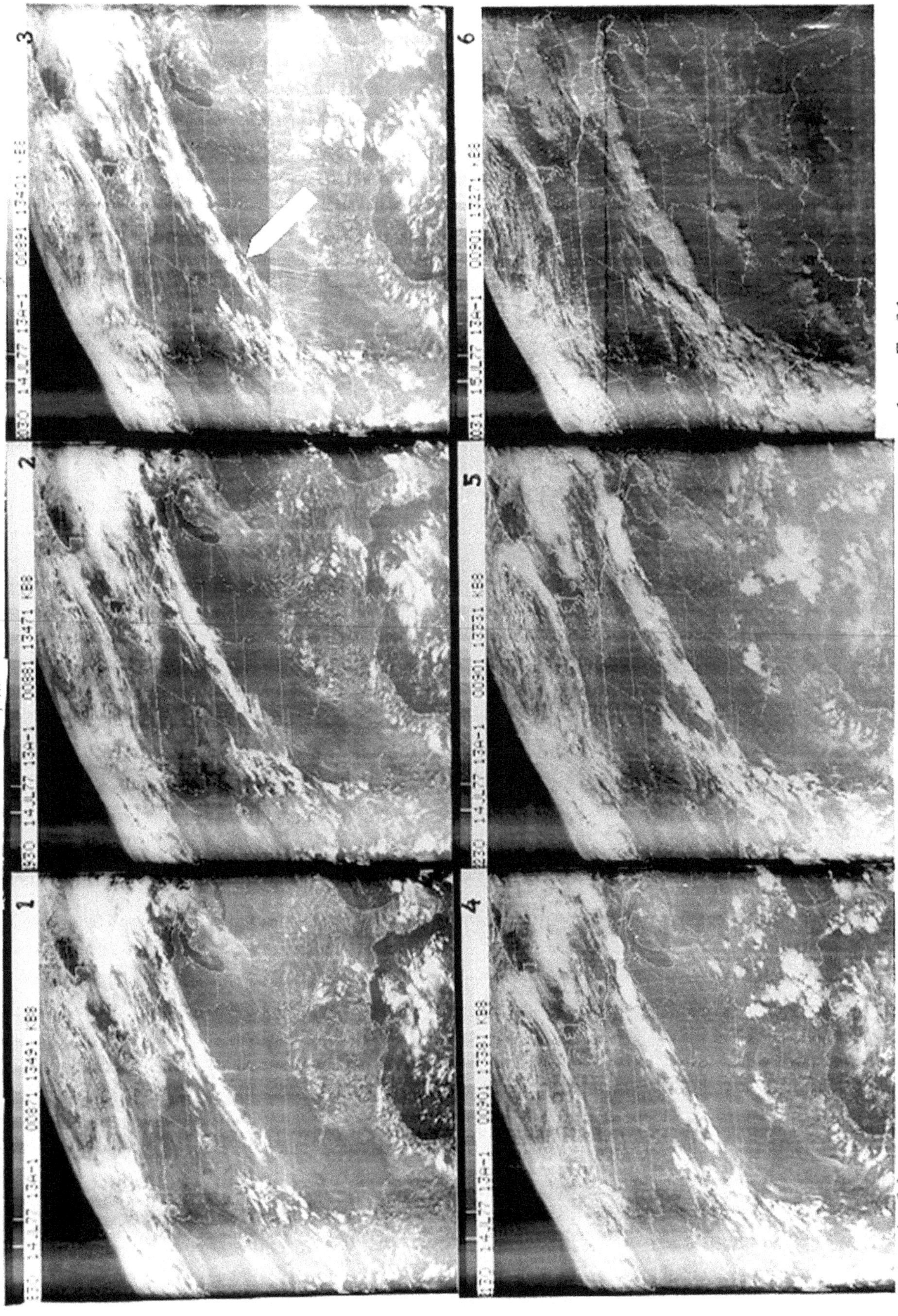

1. 1:30 pm
2. 2:30 pm
3. 3:30 pm
4. 4:30 pm
5. 5:30 pm
6. 7:31 pm

Photo Sequence 8: (on following page)
SATELLITE IMAGERY — 17 July 1977

In this series of 11 satellite images, operations started at 2:00 pm, which is just after photo #5 was taken. In these infrared images, dark areas are hot spots where latent heat is being released into the atmosphere from precipitation. While no precipitation was noted from this operation, we do note a spreading of cloud cover over the western and central portions of the state occurring just after the start of operations (see arrow, photo #5). This small area of cloud cover was relatively stable before operations, but spread outward somewhat after operations started, as seen in photos #6 through #9. Operations were terminated at 6:00 pm, which is when photo #9 was taken. Shortly thereafter, we note this area of cloud cover in Kansas to thin out and loose structure in the infrared image.

1. 10:00 am

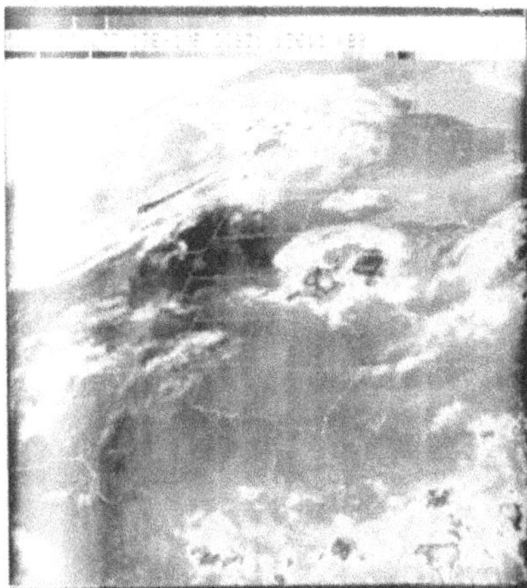

2. 11:00 am

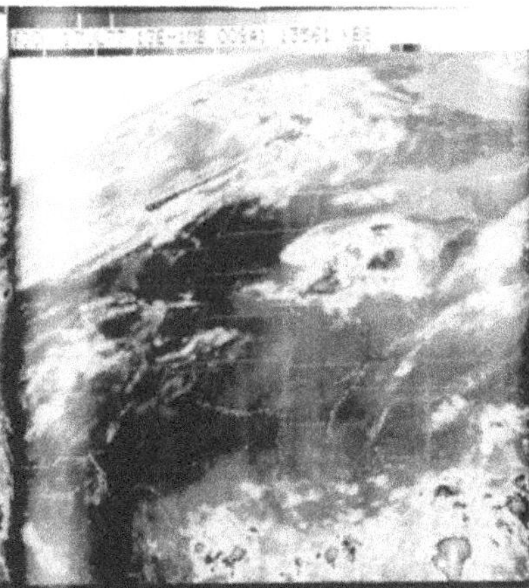

3. 12:00 pm

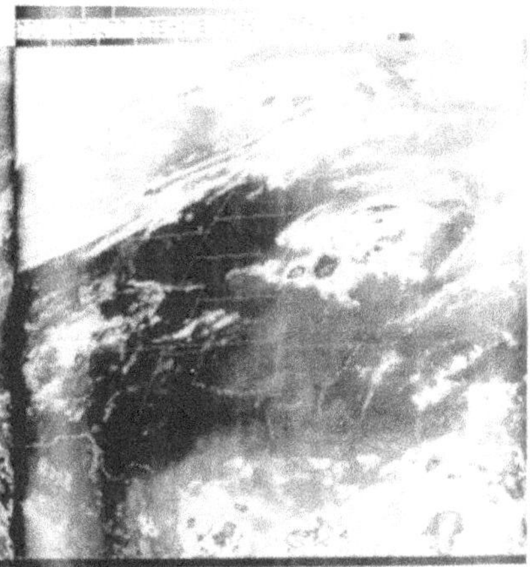

4. 1:00 pm

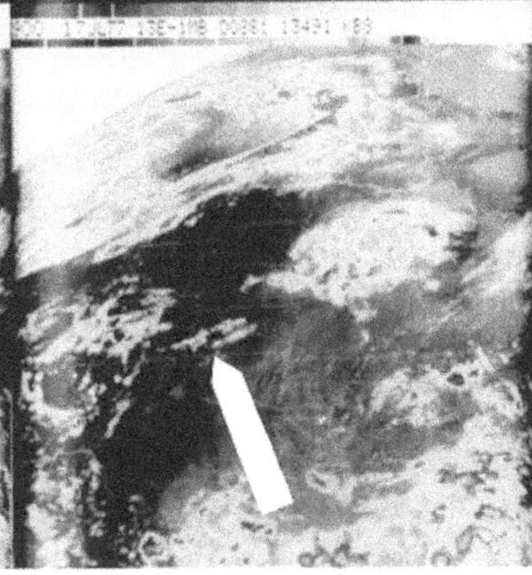

5. 2:00 pm

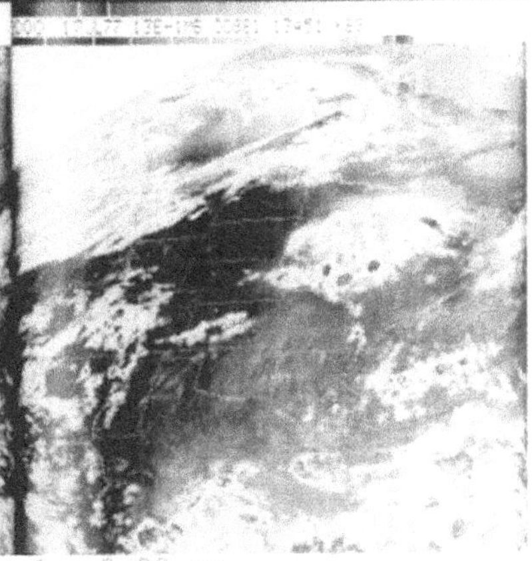

6. 3:00 pm

7. 4:00 pm

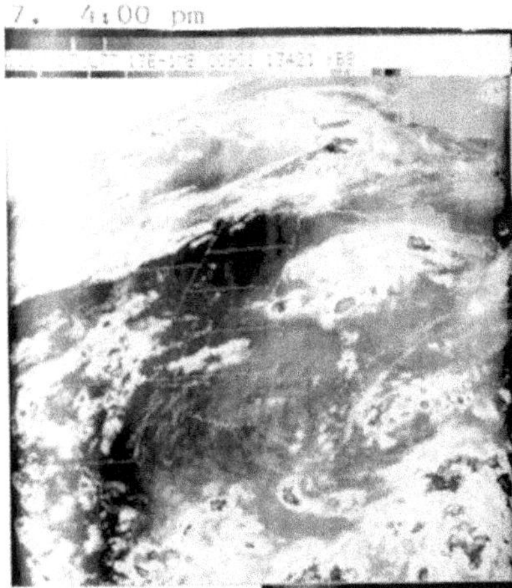

8. 5:00 pm

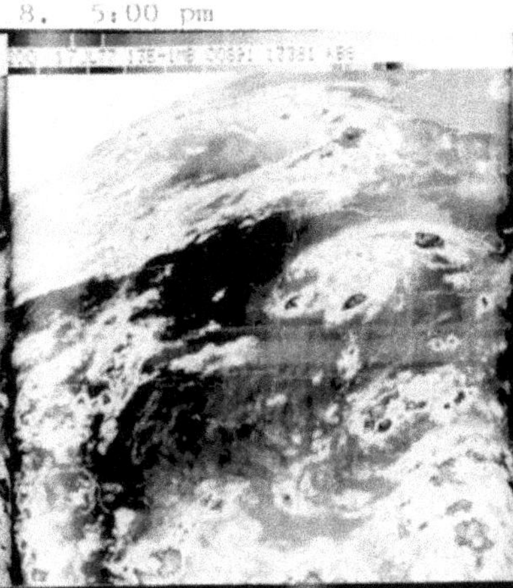

9. 6:00 pm

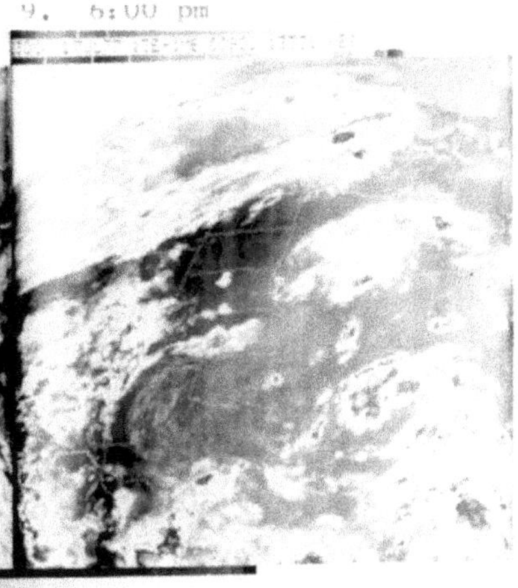

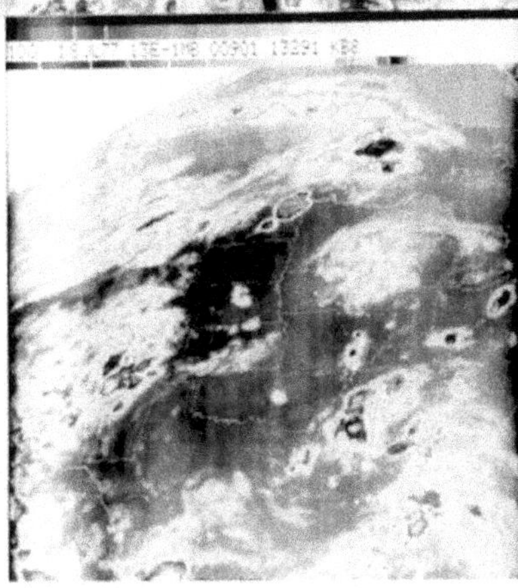

10. 8:00 pm

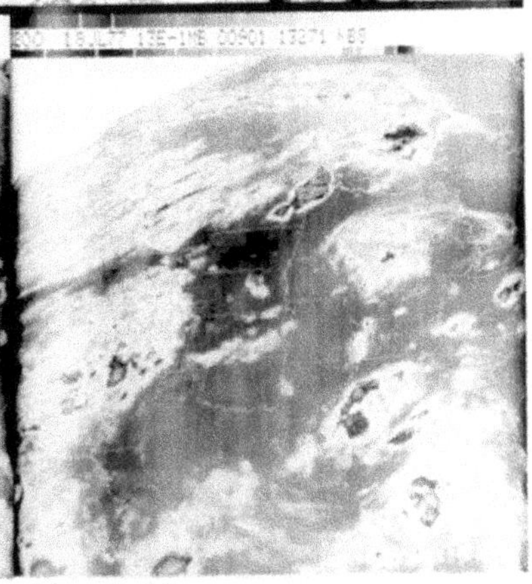

11. 9:00 pm

Photo Sequence 9: (on following page)
SATELLITE IMAGERY — 30 July 1977

In this series of 12 satellite photos, operations started at 2:00 pm, which is when photo #5 was taken. One hour later, in photo #6, a convective thunderstorm cell has blossomed about 100 miles to the west of the operational site. (These photos show the area from central Texas north through Oklahoma, Kansas, Nebraska and South Dakota) This convective storm cell continues to intensify, as seen in photos #7 - #11. Operations were terminated at 8:00 pm, which is when photo #11 was taken; by photo #12 the thunderstorm cell starts to move to the S.E., out of the area of operations.

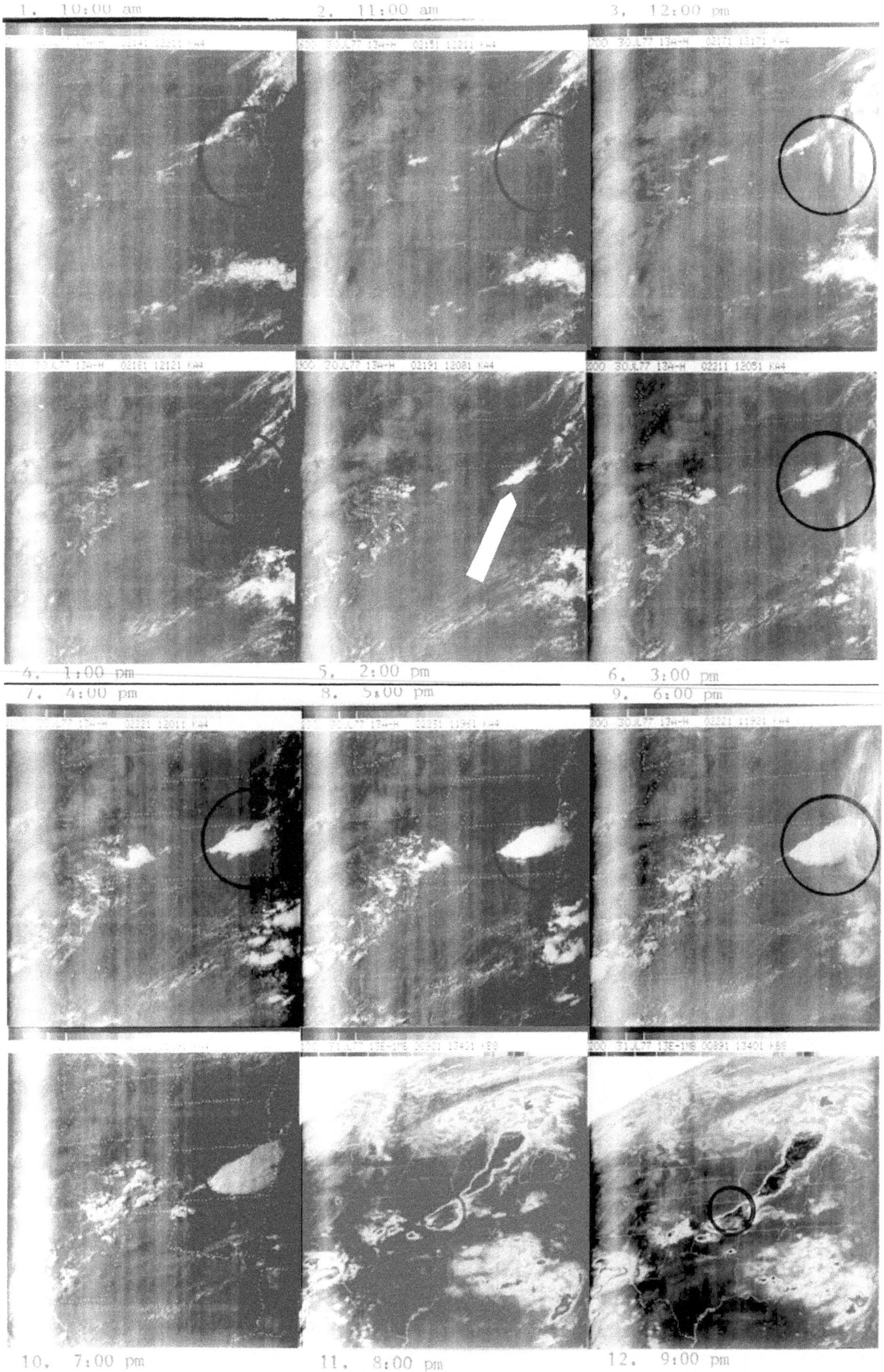
1. 10:00 am
2. 11:00 am
3. 12:00 pm
4. 1:00 pm
5. 2:00 pm
6. 3:00 pm
7. 4:00 pm
8. 5:00 pm
9. 6:00 pm
10. 7:00 pm
11. 8:00 pm
12. 9:00 pm

Photo Sequence 10: (on following page)
SATELLITE IMAGERY — 8 September 1977

This pair of satellite photos were taken at 11:00 am on the morning of and after operations, which started at 5:30 pm. While no immediate changes were noted for time of onset of operations, by the morning following operations there did occur a great organization in a low pressure center many miles to the north, with a cold front extending southward over Kansas.

1. 11:00 am

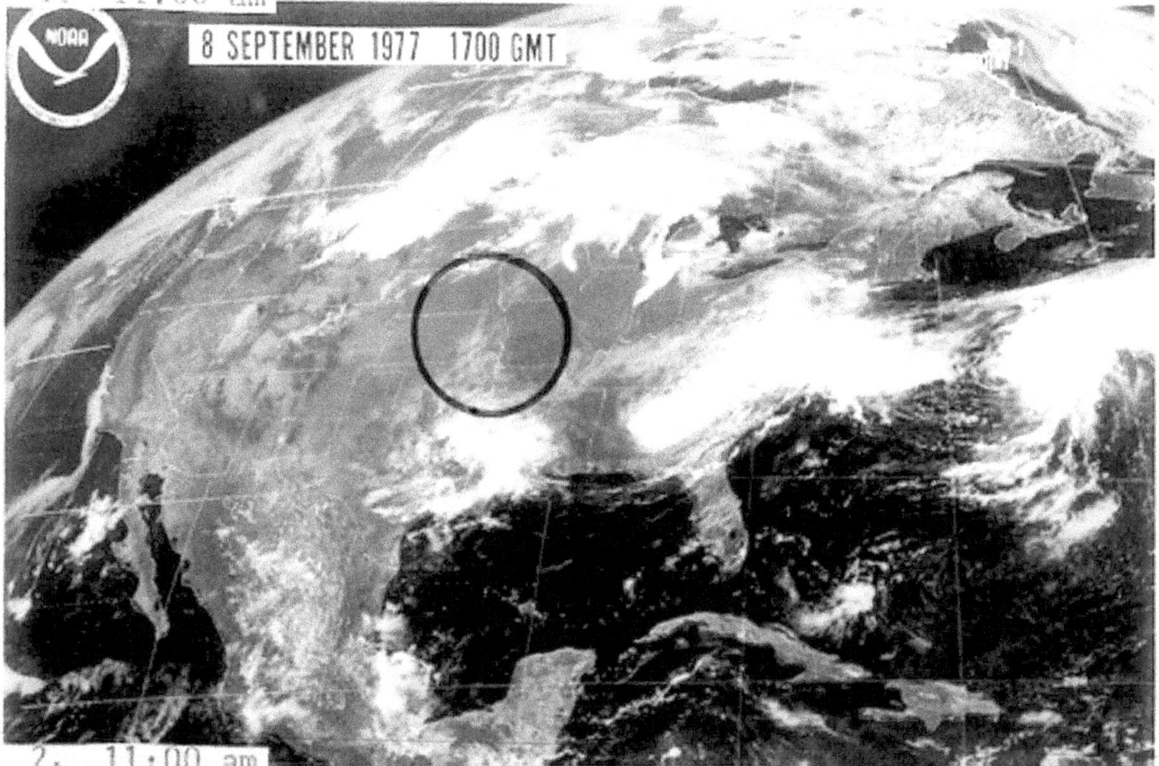

2. 11:00 am

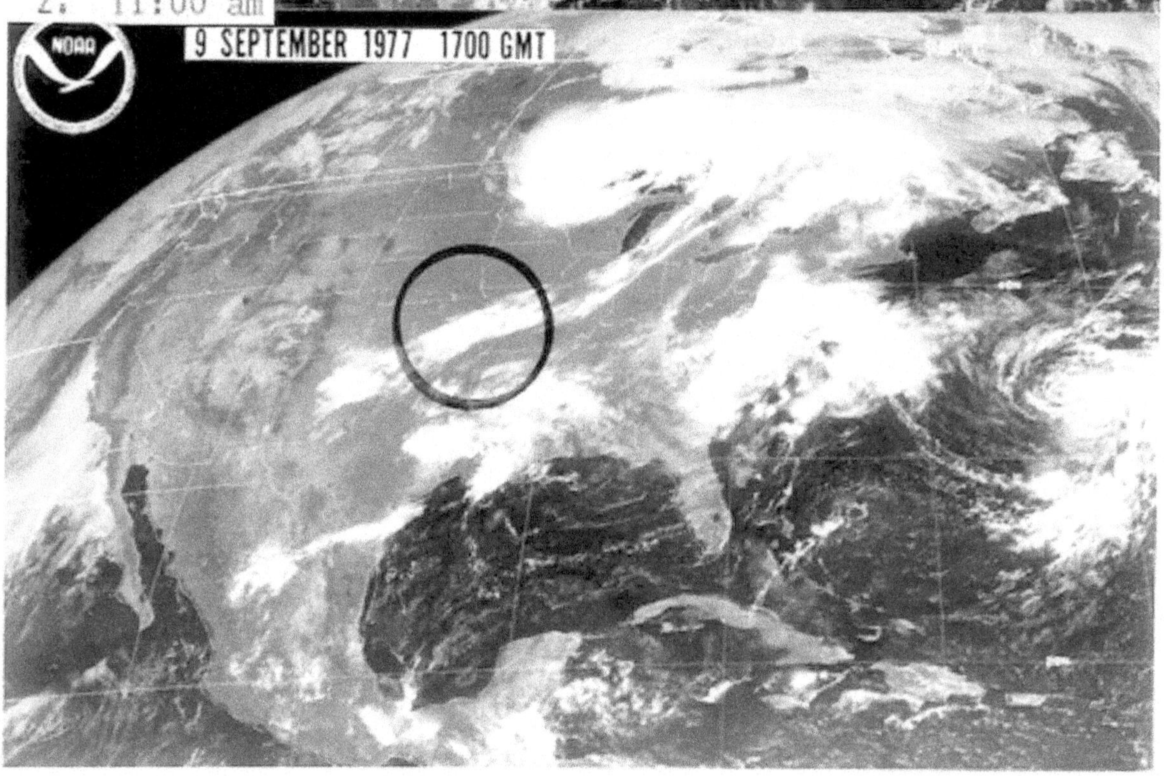

Photo Sequence 11: (on following page)
SATELLITE IMAGERY — 26 August 1978

In this series of four satellite images, operations started at 12:15 pm, which is just before photo #2. Photos #1 and #3 are visible-light imagery and photos #2 and #4 are infrared imagery. The frontal system that stretches N. - S. across Kansas shows signs of intensification by photo #3, which is about two hours after the operation started. By 7:30 pm, this frontal system had intensified greatly, as shown in photo #4 (dark area indicates infrared hot spot — release of latent heat from precipitation).

1. 12:00 pm

2. 12:30 pm

3. 2:00 pm

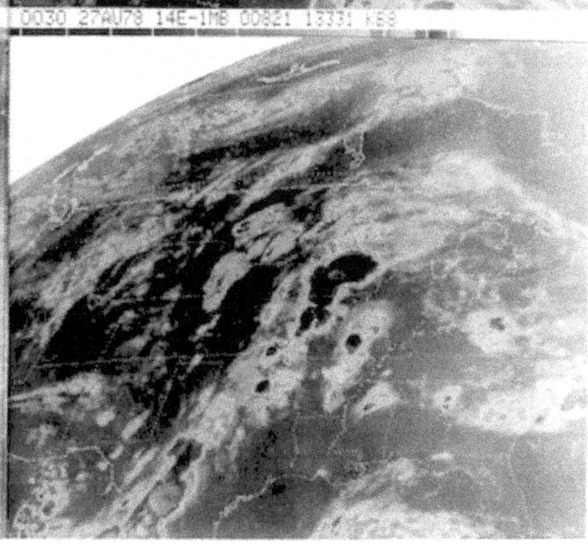

4. 7:30 pm

Photo Sequence 12: (on following page)
SATELLITE IMAGERY — 3 September 1978

In this series of six satellite photos, operations started at 12:55 pm, just after photo #1. Shortly after operations commenced, an area of cloudiness is seen to develop in the extreme S.E. corner of Kansas (see arrow, photo #2). This area of cloudiness blossoms into cumulonimbus within four hours after the start of operations, as seen in photos #3 : #5. Photo #5 is an infrared image depicting a latent-heat hot spot in S.E. Kansas at 9:30 pm. This area migrated northward during the night and dissipated slowly, as seen in photo #6, which was taken at 3:00 am on the morning following cloudbusting operations.

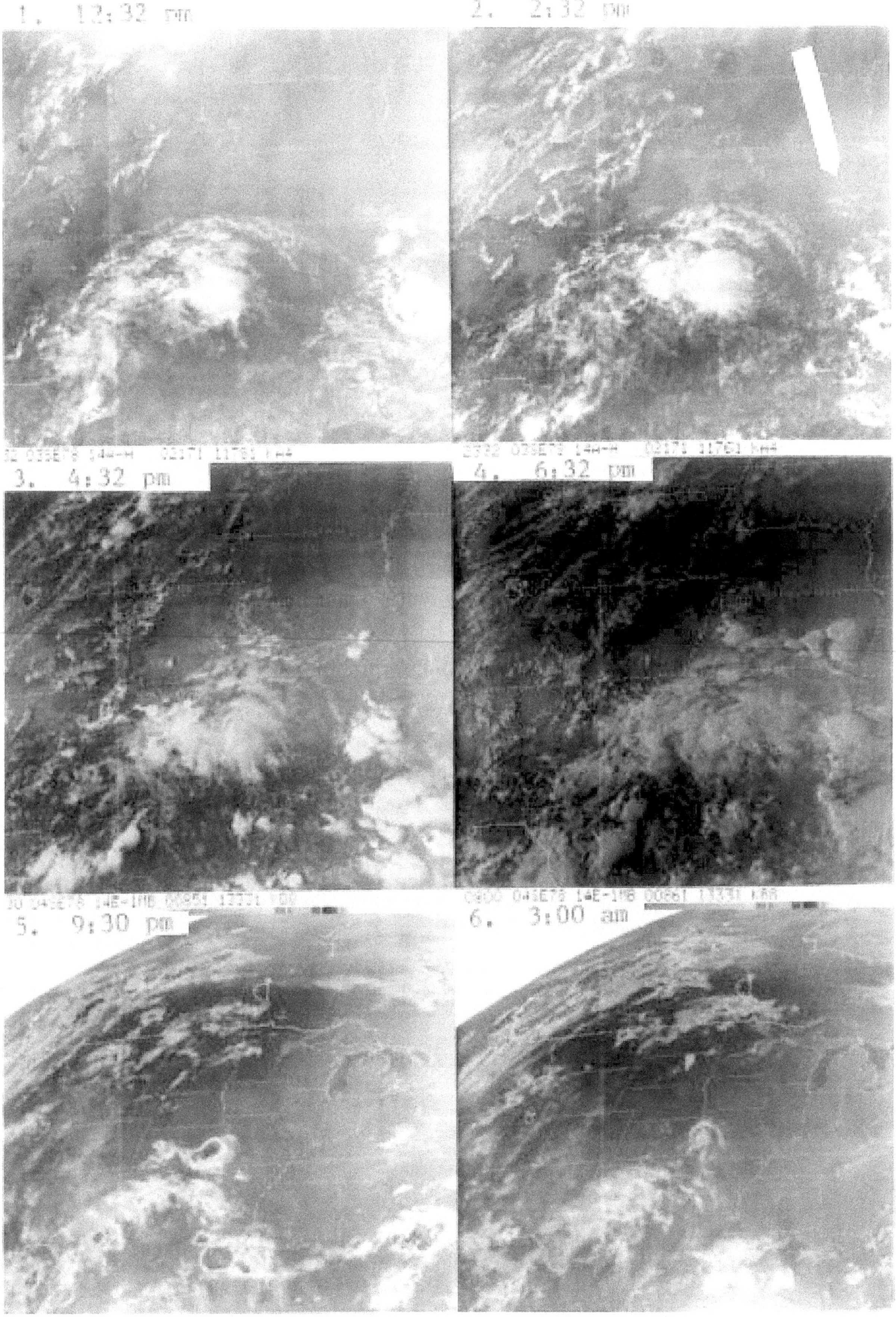
1. 12:32 pm
2. 2:32 pm
3. 4:32 pm
4. 6:32 pm
5. 9:30 pm
6. 3:00 am

Photo Sequence 13: (on following page)
SATELLITE IMAGERY — 10 September 1978

The first of this pair of satellite photos was taken 3:00 pm, just as cloudbusting operations commenced. Scattered cumulus with some convective cell activity was present over the area. The second photo shows conditions at 11:00 am the following morning. The cloudiness over eastern and S.E. Kansas intensified into what appears to be thick stratus.

1. 3:00 pm

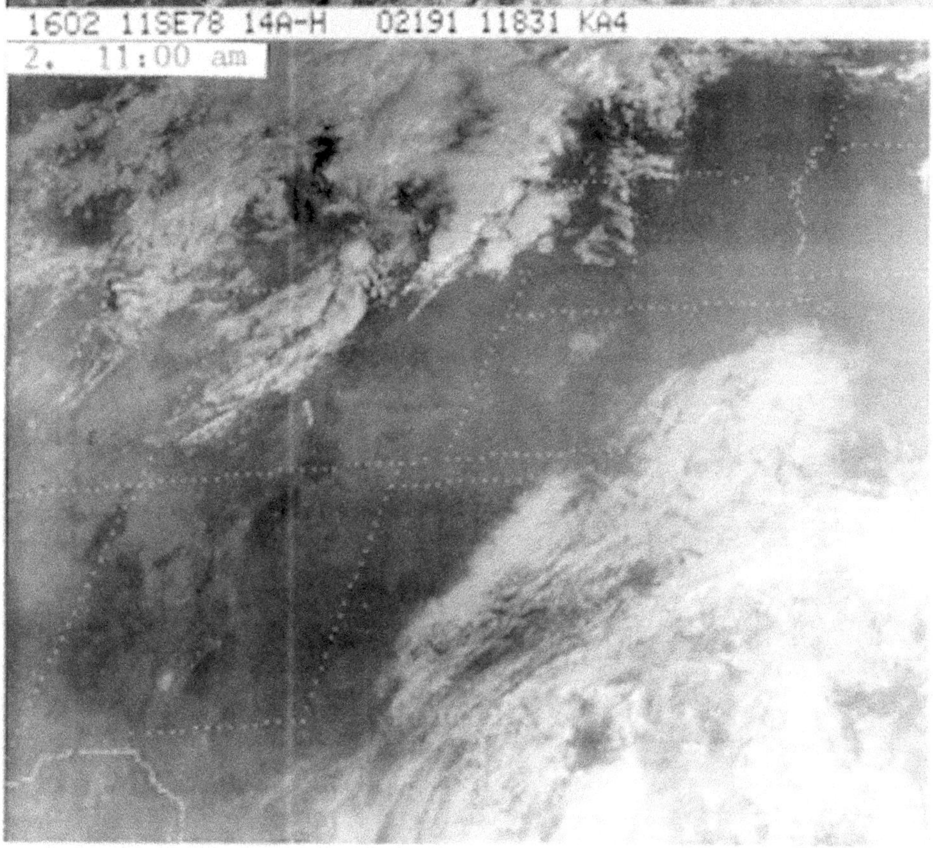
1602 11SE78 14A-H 02191 11831 KA4
2. 11:00 am

Photo Sequence 14: (on following page)
SATELLITE IMAGERY — 17 September 1978

The first of this pair of infrared satellite images shows conditions at 10:30 am on the day of cloudbuster operations, which started at 12:20 pm. An area of intense rainfall was located on the Kansas-Nebraska border. By 8:00 pm of the same day this area seems to have greatly intensified, bringing precipitation from Texas northward to Iowa and across Illinois and Indiana.

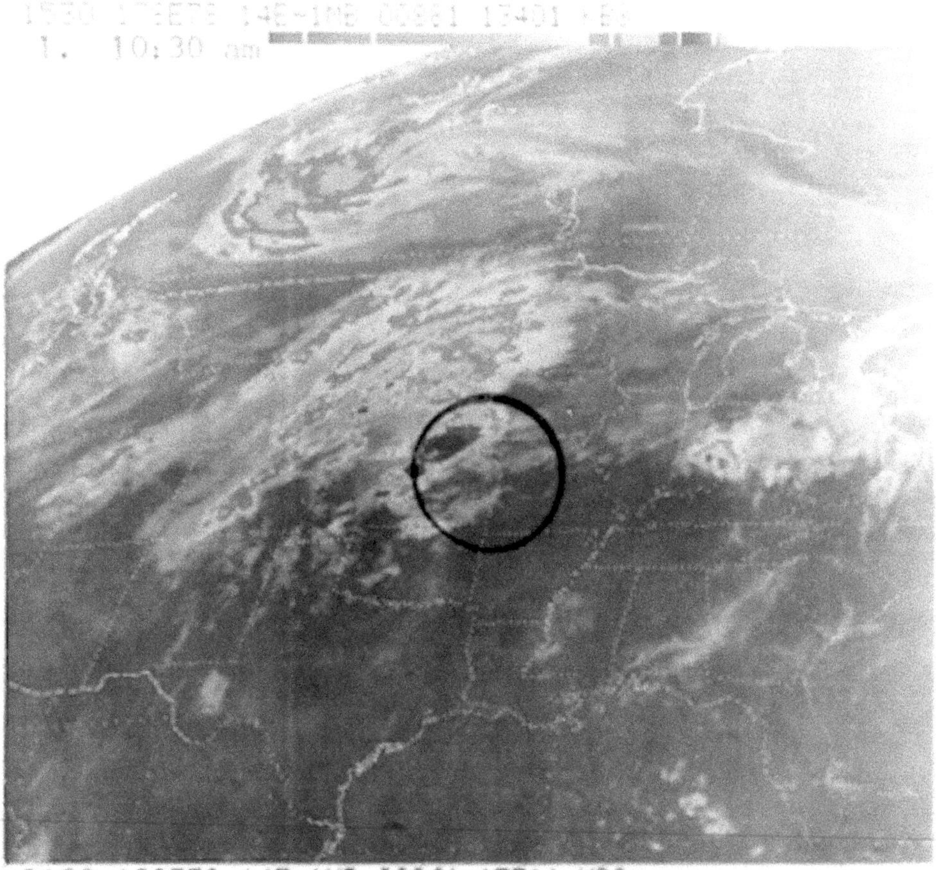
1. 10:30 am

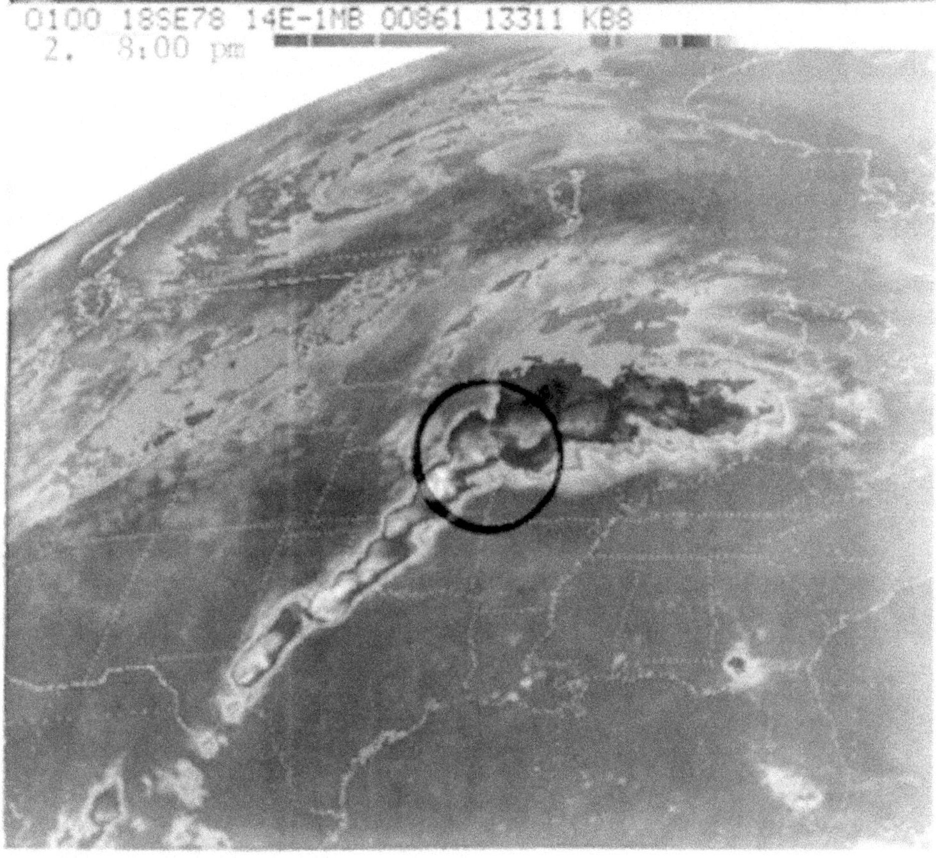
0100 18SE78 14E-1MB 00861 13311 KB8
2. 8:00 pm

Photo Sequence 15: (on following page)
SATELLITE IMAGERY — 24 September 1978

In this series of six satellite photos, operations started at 12:40 pm, just after photo #2. An area of cloudiness to the s.w. had been advancing slowly into the area before operations commenced. The area seems to aggregate somewhat, in photo #4, which was taken at 9:00 pm on the day of operations. The most intense area of activity in this storm system sweeps northeast during the night crossing over N.E. Kansas (see arrow, photo #6).

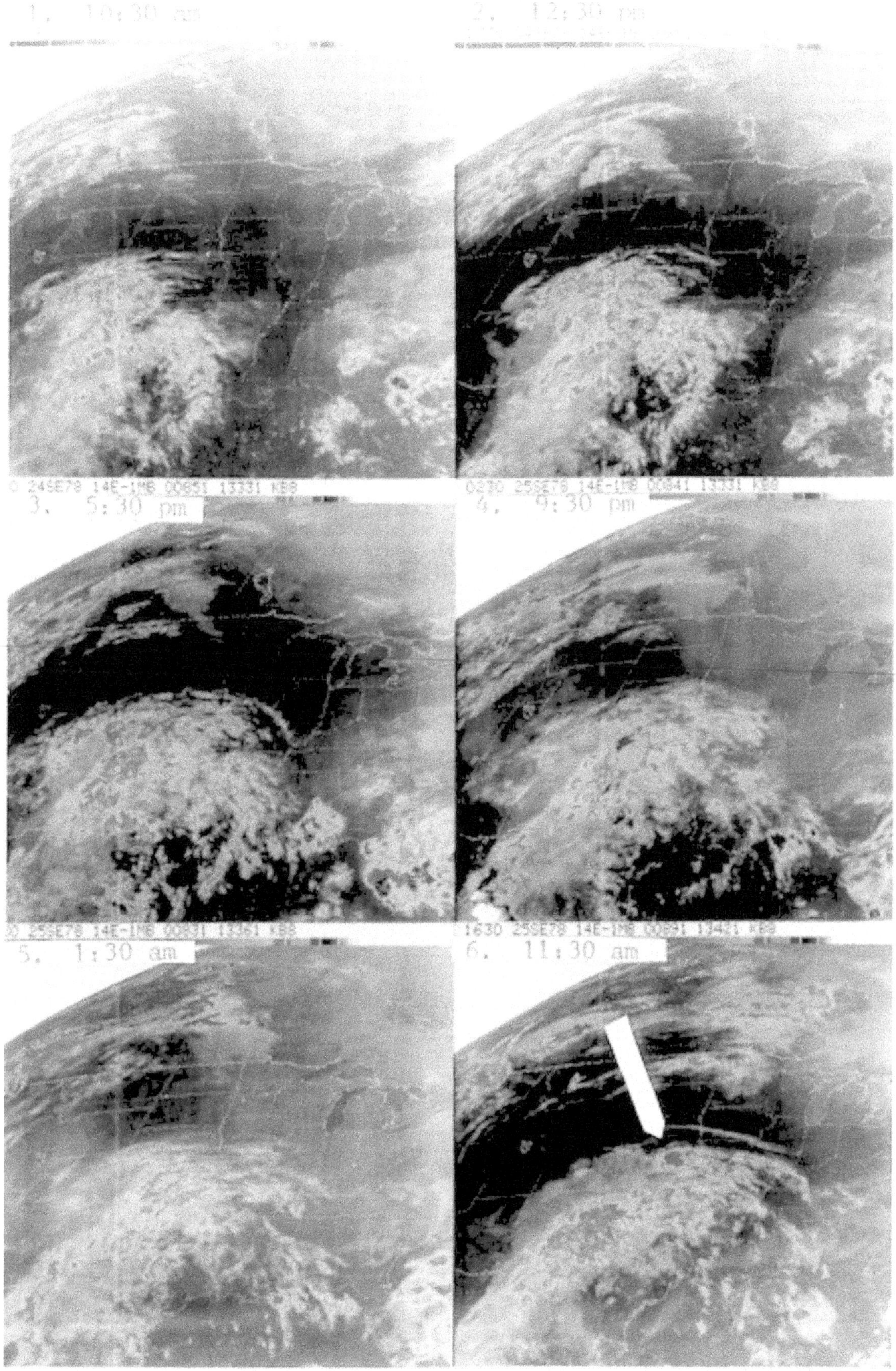
1. 10:30 am
2. 12:30 pm
3. 5:30 pm
4. 9:30 pm
5. 1:30 am
6. 11:30 am

Photo Sequence 16: (on following page)
SATELLITE IMAGERY 1 October 1978

The first of this pair of satellite photos was taken at 12:02 pm, one half hour before cloudbusting operations commenced. A band of cumulus stretching N. - S. was over the Kansas Flint Hills area at this time. The second of the two satellite photos was taken at 4:02 pm on the following day, revealing cloud cover and convective activity across N.E. Kansas.

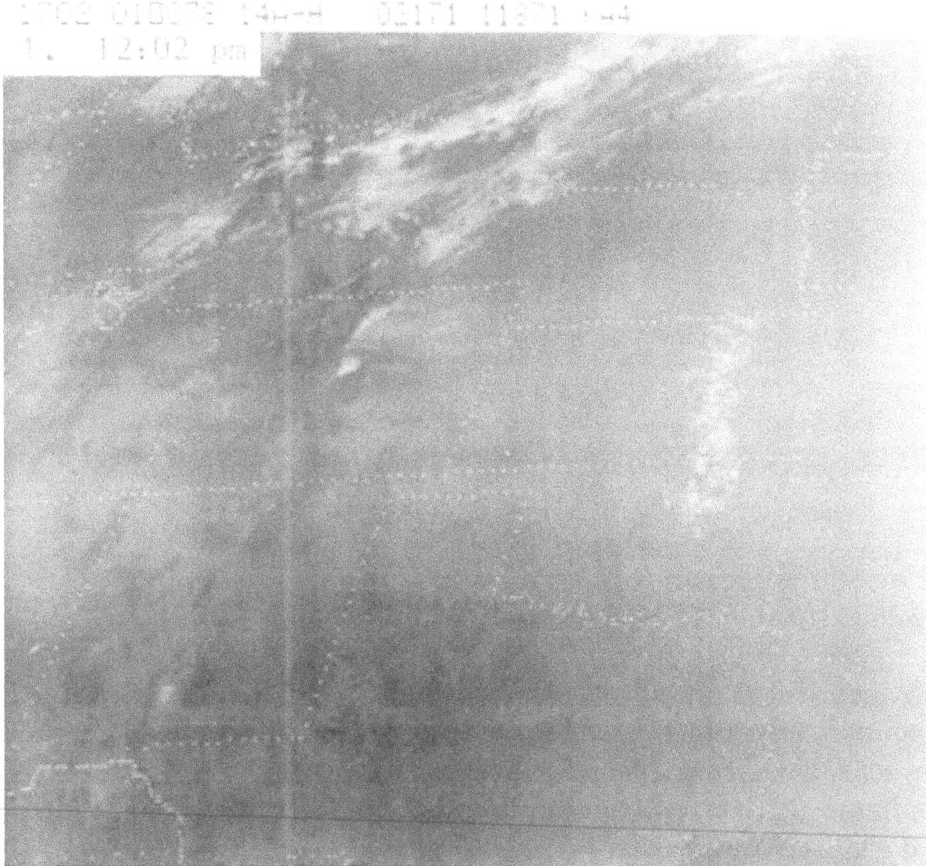
1. 12:02 pm

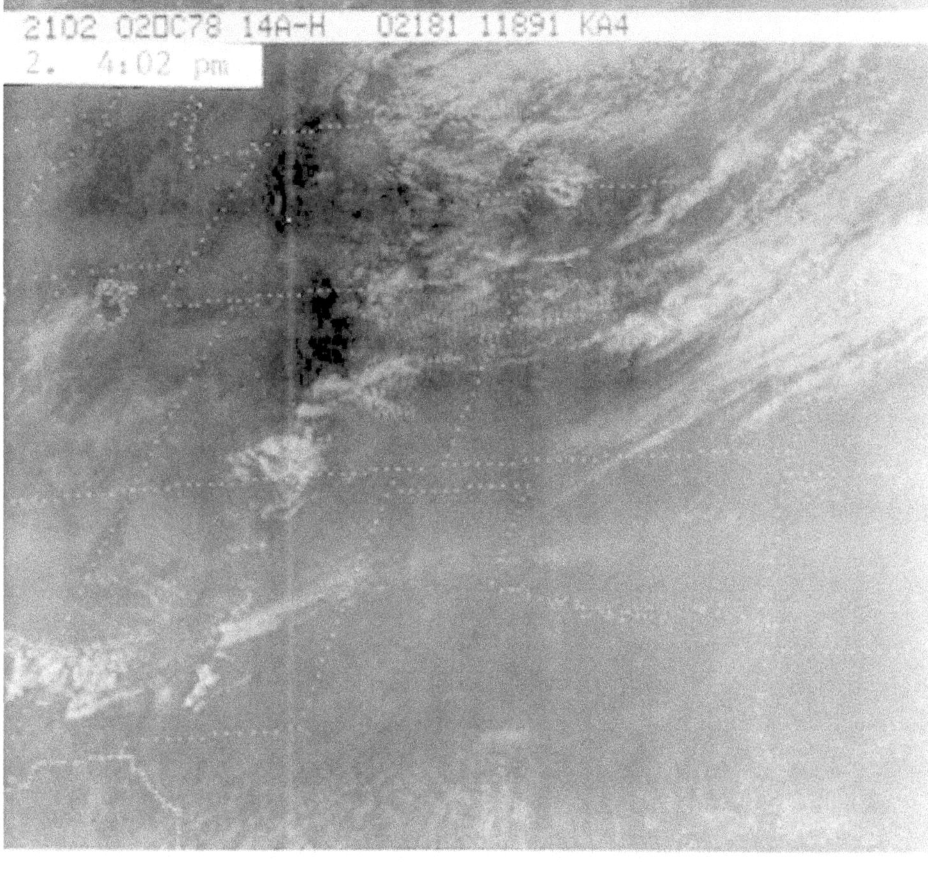
2102 020C78 14A-H 02181 11891 KA4
2. 4:02 pm

Photo Sequence 17: (on following page)
SATELLITE IMAGERY — 8 October 1978

In this sequence of six satellite images, cloudbuster operations started at 12:35 pm, 1/2 hour after photo #1 was taken. In this sequence, we see a frontal system which sweeps down from the Northeast during the night. The southernmost center of heavy activity, as seen in photos #5 and #6, intensifies while passing eastward across N.E. Kansas.

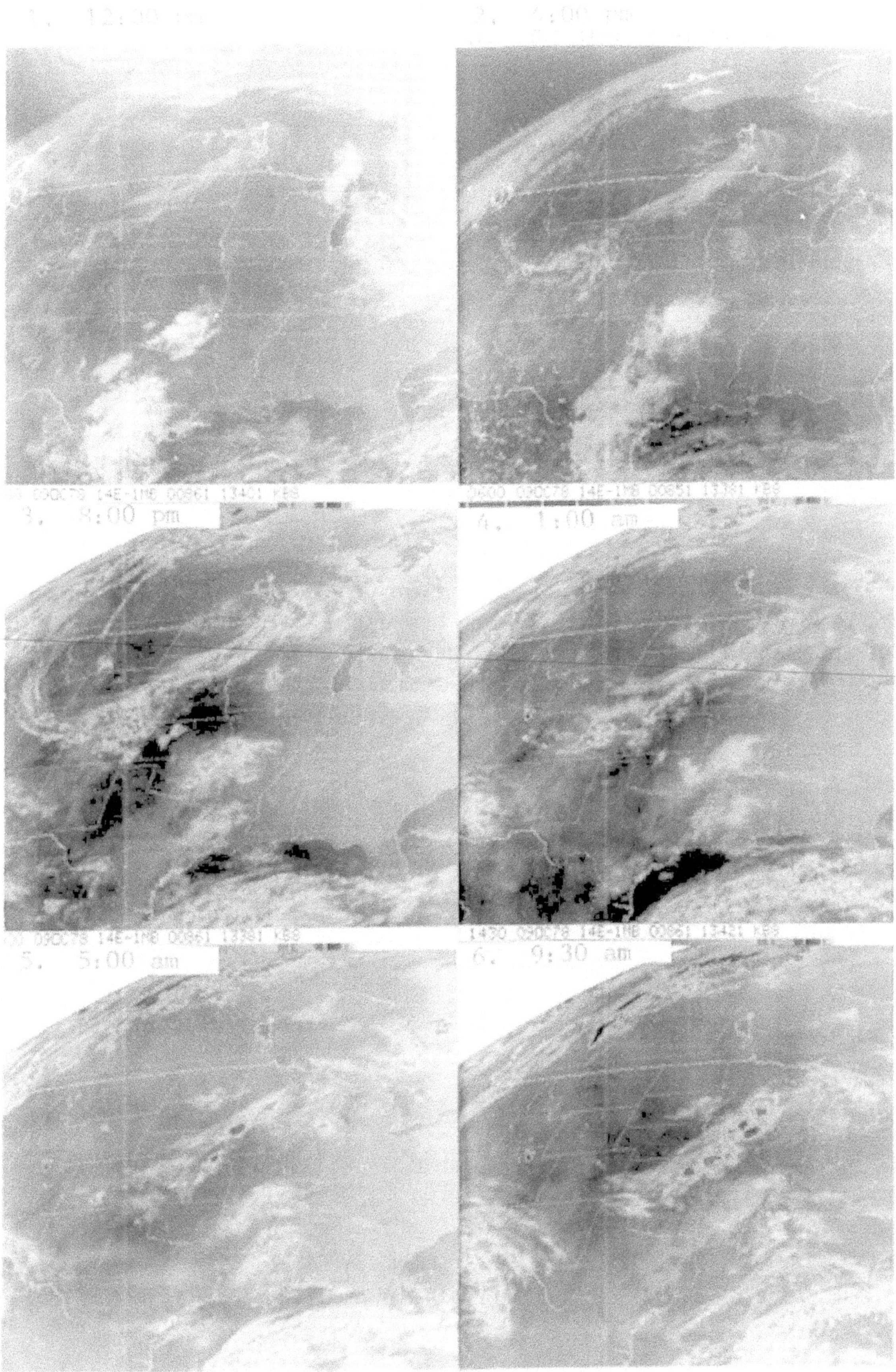
1. 12:30 pm
2. 4:00 pm
3. 8:00 pm
4. 1:00 am
5. 5:00 am
6. 9:30 am

Photo Sequence 18: (on following page)
SATELLITE IMAGERY — 15 October 1978

This satellite photo was taken at noon on the day of operations, one half hour before operations, which began one half hour later. No clouds were visible from the Mississippi to the Pacific Ocean. This operation did not result in any significant in weather data (see Figure 43). However, an unexpected frontal passage did occur (see weather maps 15-16 October 1978 in Appendix IV.)

1. 12:00 pm

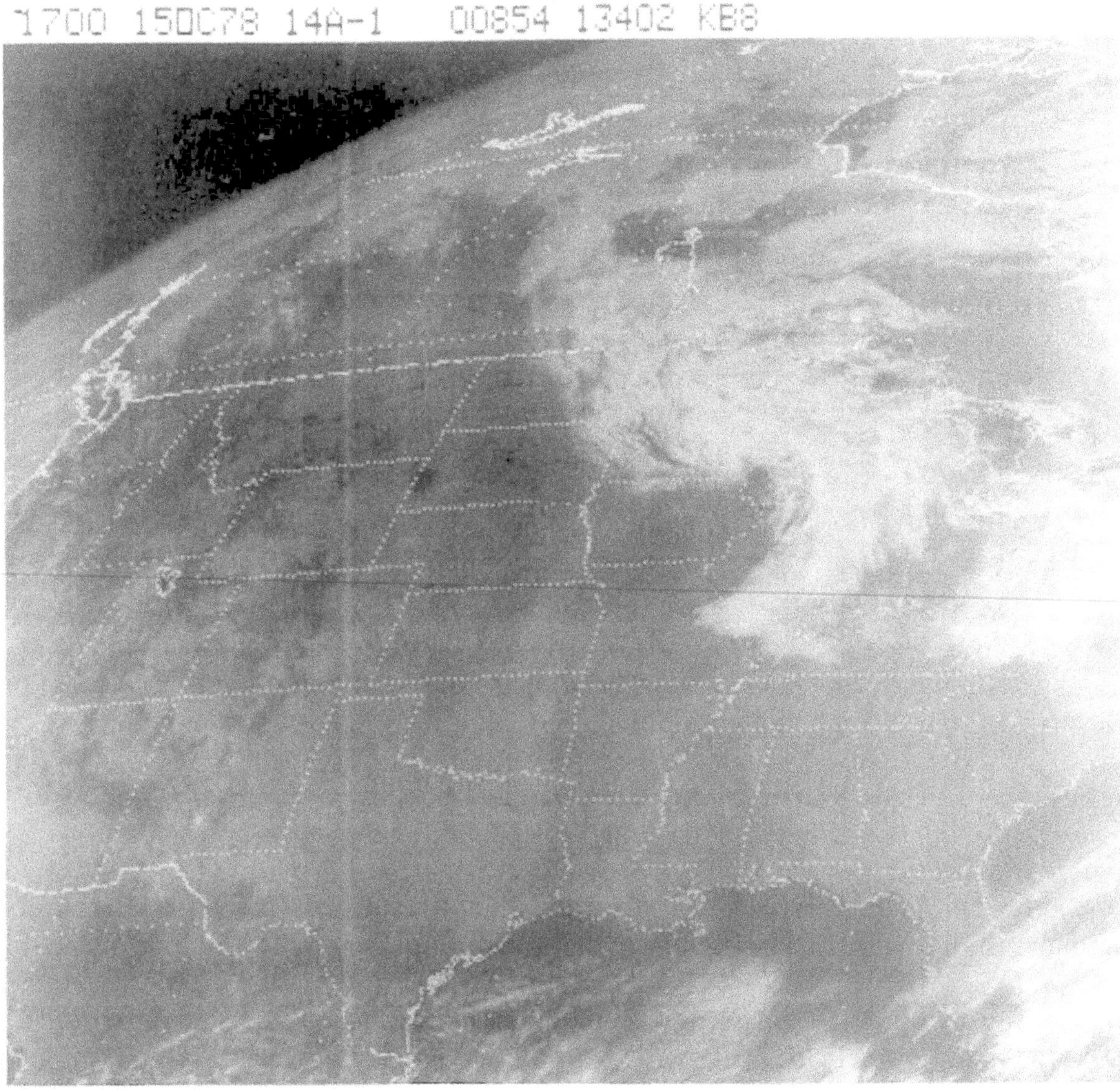

Table II, below, summarizes the 12 operations with regard to date of test and satellite image observations from Photo Sequences 7 - 18.

Table II: SATELLITE IMAGE OBSERVATIONS

	Effect Observed			
Operational Date	Immediate effect 1	Within a few hours 2	Within 24 hours 3	None observed 4
14 July 77	X			
17 July 77	X			
30 July 77	X			
8 Sept 77			X	
26 Aug 78	X			
3 Sept 78		X		
10 Sept 78			X	
17 Sept 78		X		
24 Sept 78			X	
1 Oct 78			X	
8 Oct 78			X	
15 Oct 78				X

From Table II, we may note a general trend from "immediate" associations in July and August toward later and later associations for September and October. The trend fits well with two possibilities:

A. Perhaps, the cloudbuster intensifies preexisting weather conditions or acts in a way to duplicate storm patterns that are normal for a specific geographic region and season. For example, in July and August local "spontaneous" convective thunderstorm cells are prevalent in the test area. But September and October trend toward the long wave cyclone and frontal activity predominates.

B. Possibly, by quirk of chance, these cloudbuster tests were performed exactly prior to when natural cloud blossoming and frontal activity would have occurred anyway. However, in order to negate this possibility and develop the thesis, we must be able to demonstrate that the intensity of weather for days before and after the test date were largely absent of similar buildup in activity. Assuming the cloudbuster works, we should also be able to show that most of the rainfall on the day of operations occurred after the cloudbuster was activated, and that this rainfall was greater than for several days previous to and after the date of operations.

Careful scrutiny of these satellite images further reveals that, should the cloudbuster be proved the triggering agent, such effects may extend much farther than Kansas alone. Indeed, frontal activity associated with some of these operational dates stretches for several thousand miles. While such a long distance effect may make the technique seem more of an enigma, such effect is wholly possible within the theory of operation developed by Reich (see Appendix I and III).

Data Analysis

Figure 31 is a summary of weather data for the 12 precipitation enhancement operations combined, graphing hourly and daily precipitation and % cloud cover over the state of Kansas for a weekly period centered at the onset of cloudbuster operations. Figures 32 - 43 set forth similar data for each of the 12 operations separately.

Figure 31: WEEKLY SUMMARY -- TWELVE PRECIPITATION ENHANCEMENT OPERATIONS

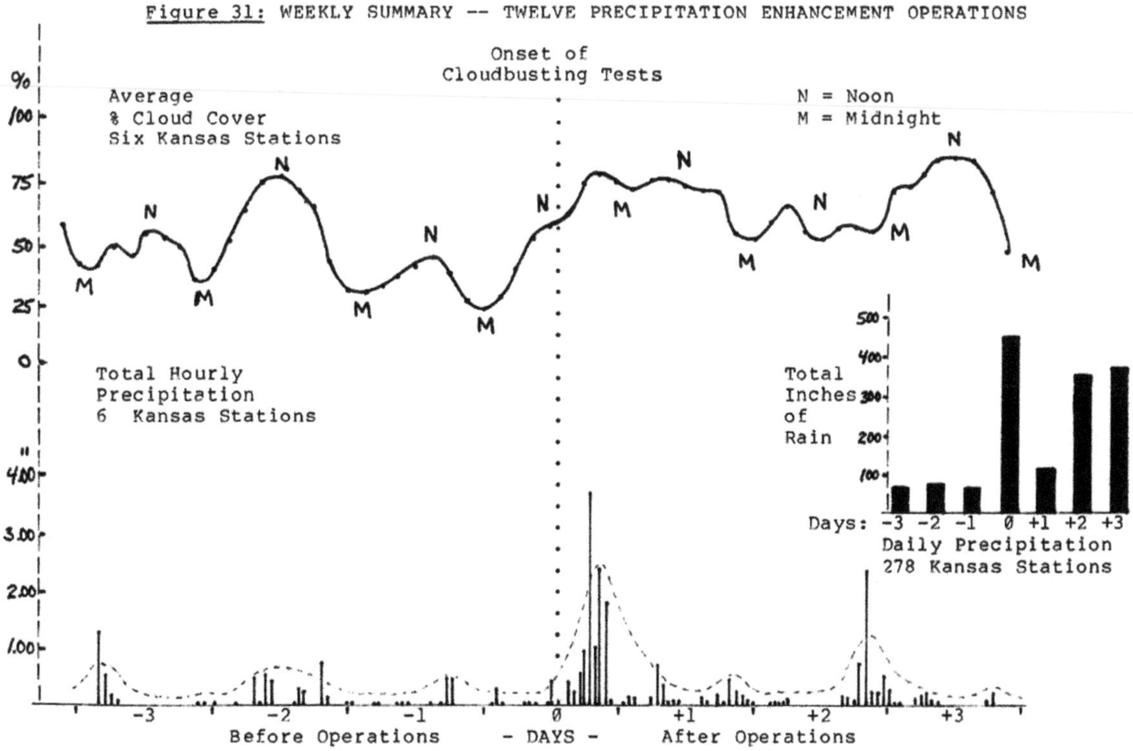

Figure 32: PRECIPITATION ENHANCEMENT OPERATION -- 14 July 1977

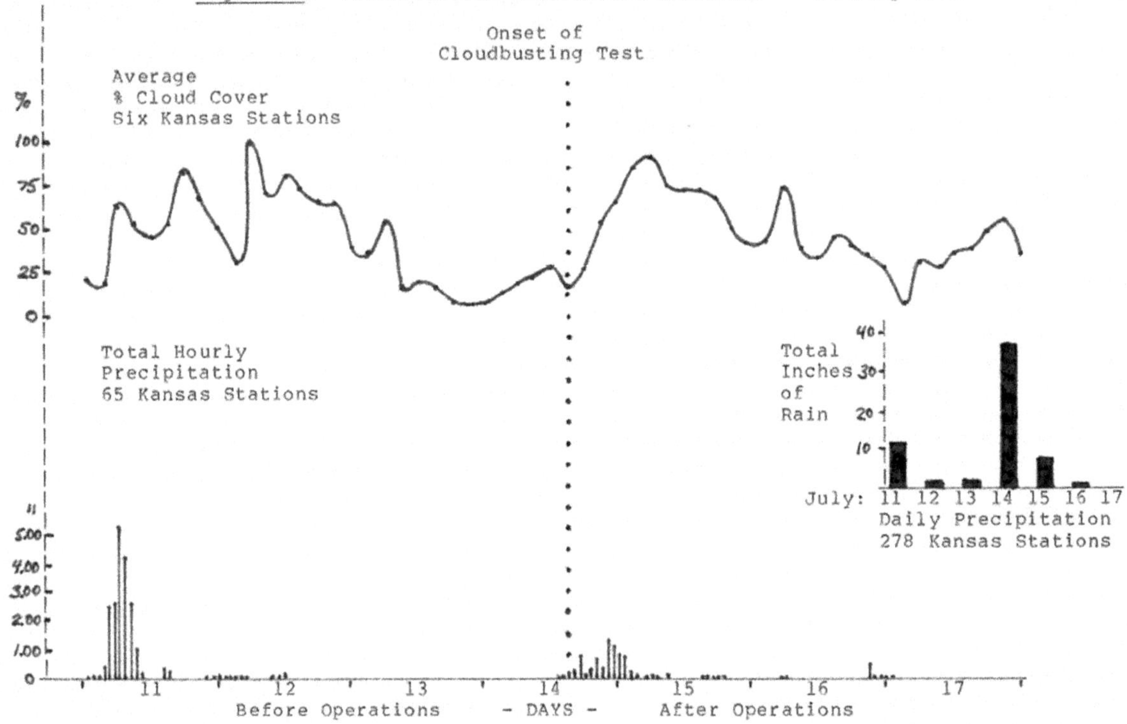

Figure 33: PRECIPITATION ENHANCEMENT OPERATION -- 17 July 1977

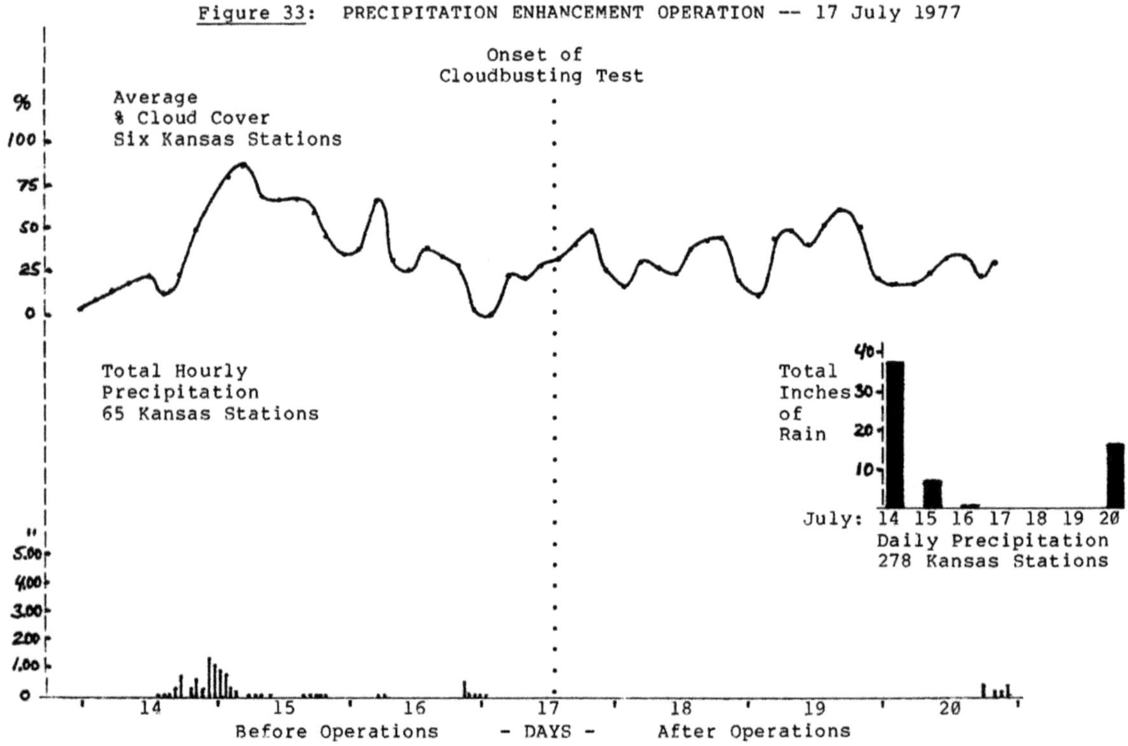

Figure 34: PRECIPITATION ENHANCEMENT OPERATION -- 30 July 1977

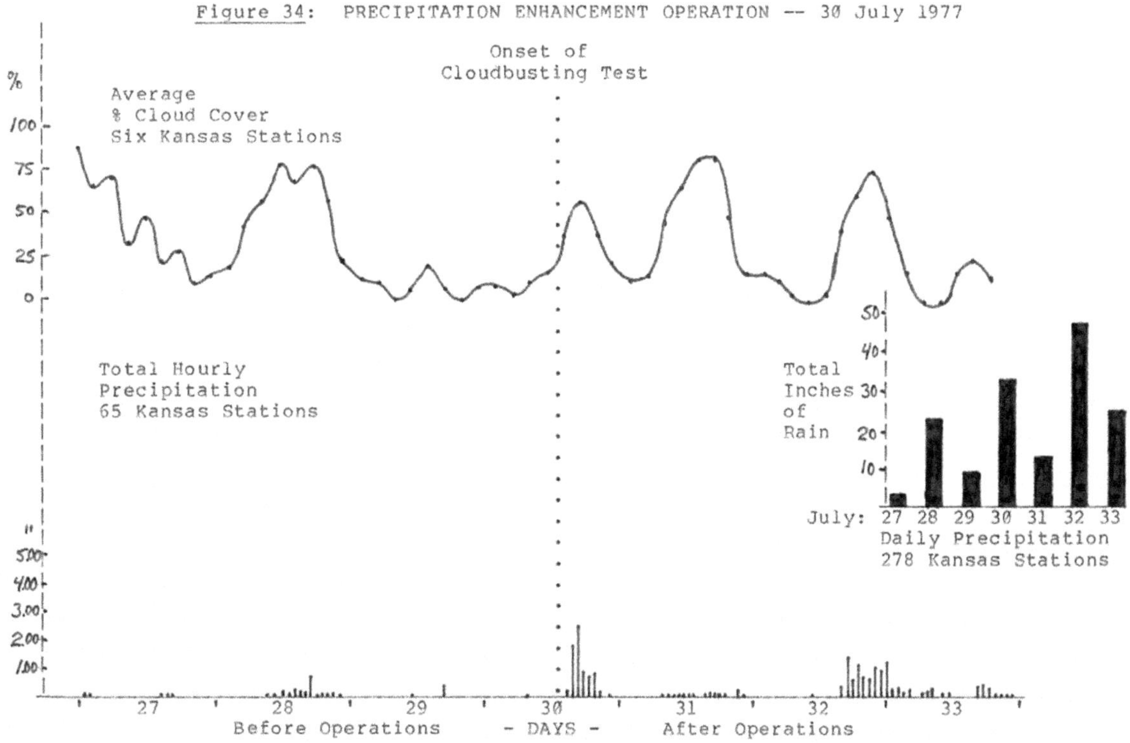

Figure 35: PRECIPITATION ENHANCEMENT OPERATION -- 8 September 1977

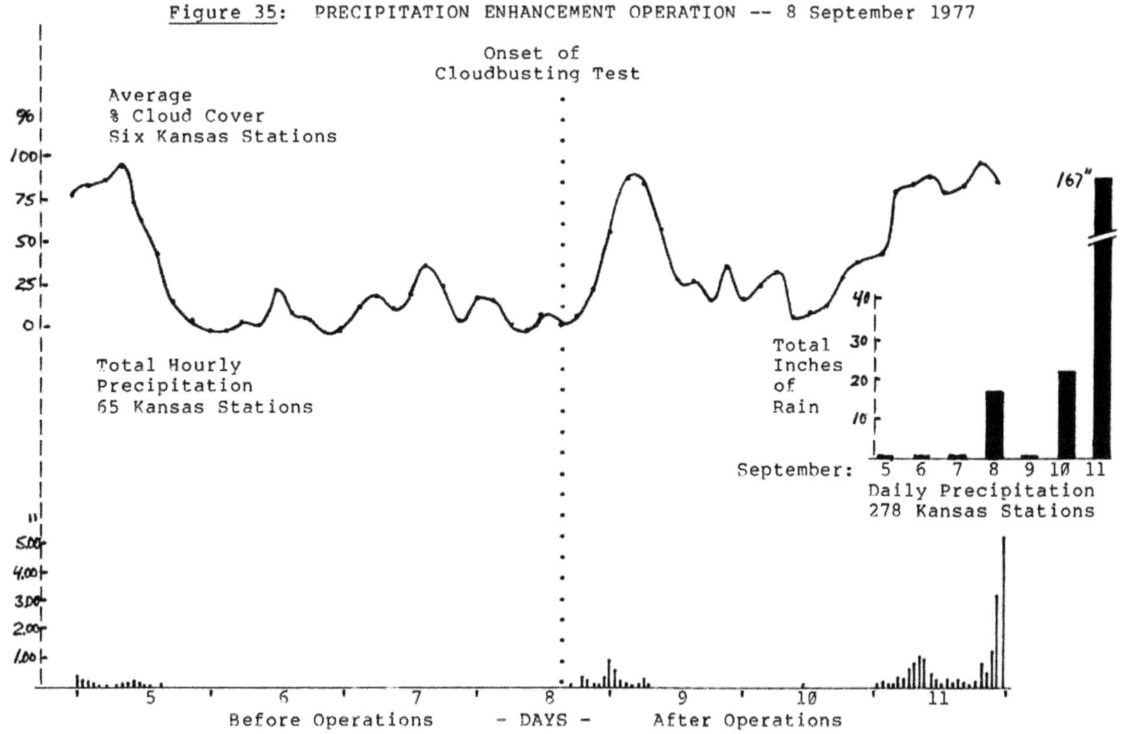

Figure 36: PRECIPITATION ENHANCEMENT OPERATION -- 26 August 1978

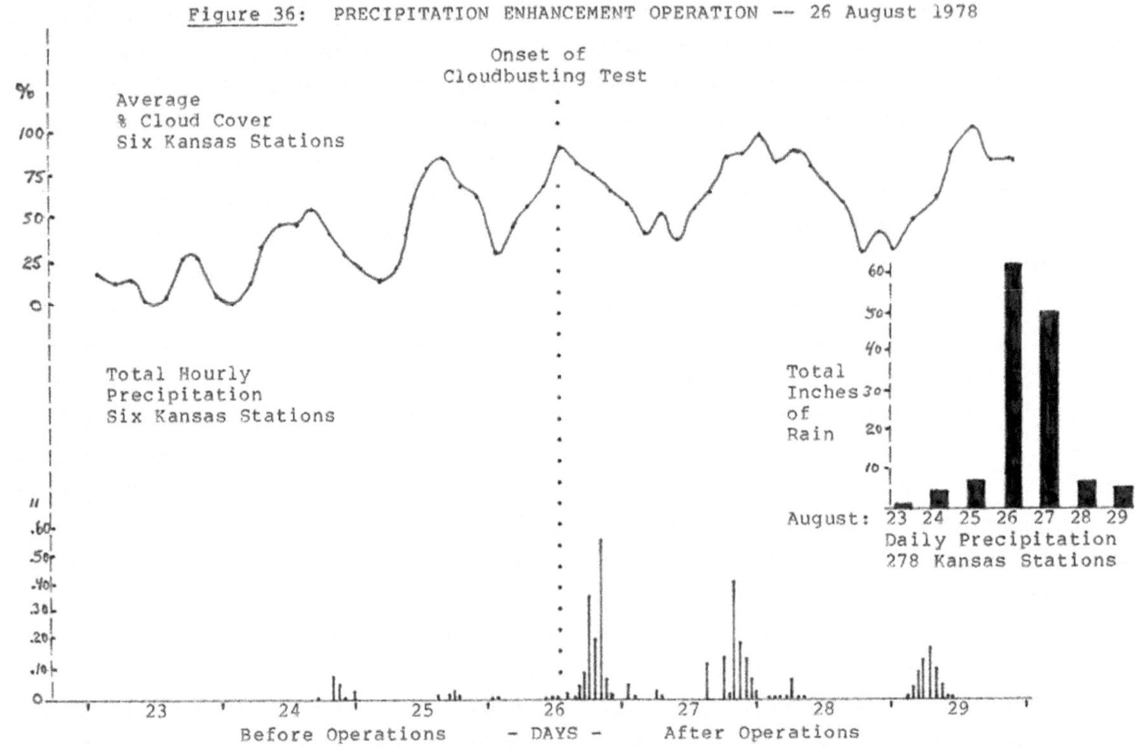

Figure 37: PRECIPITATION ENHANCEMENT OPERATION -- 3 September 1978

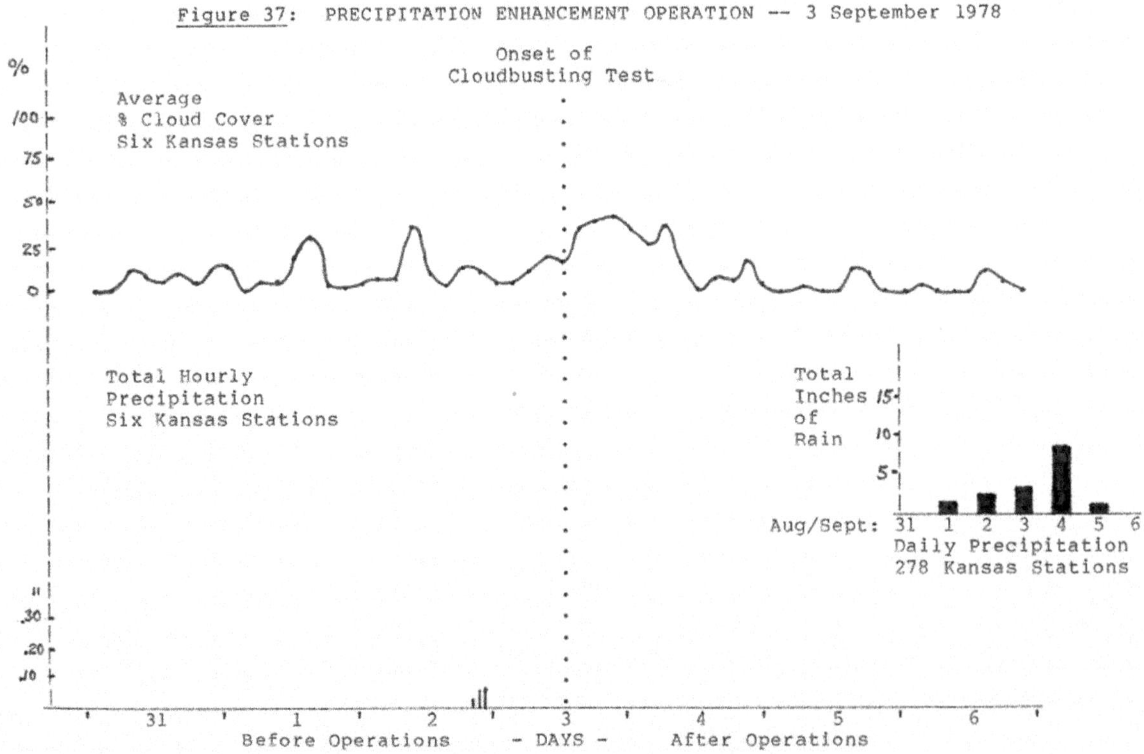

Figure 38: PRECIPITATION ENHANCEMENT OPERATION -- 10 September 1978

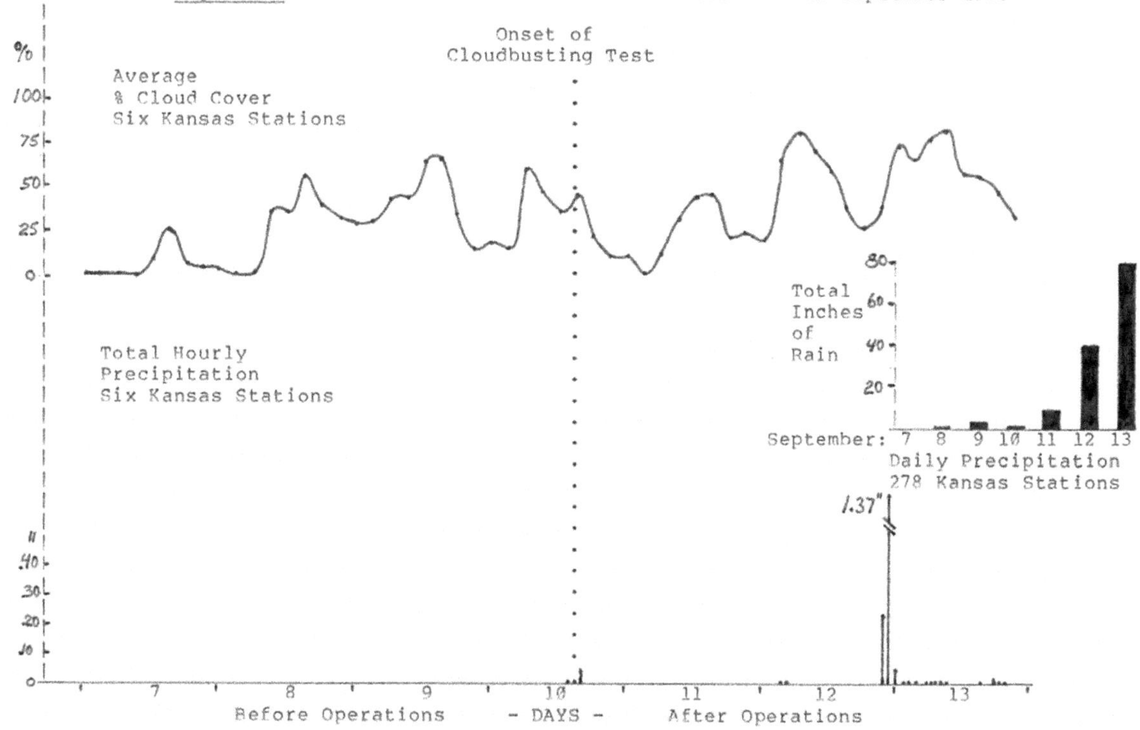

Figure 39: PRECIPITATION ENHANCEMENT OPERATION -- 17 September 1978

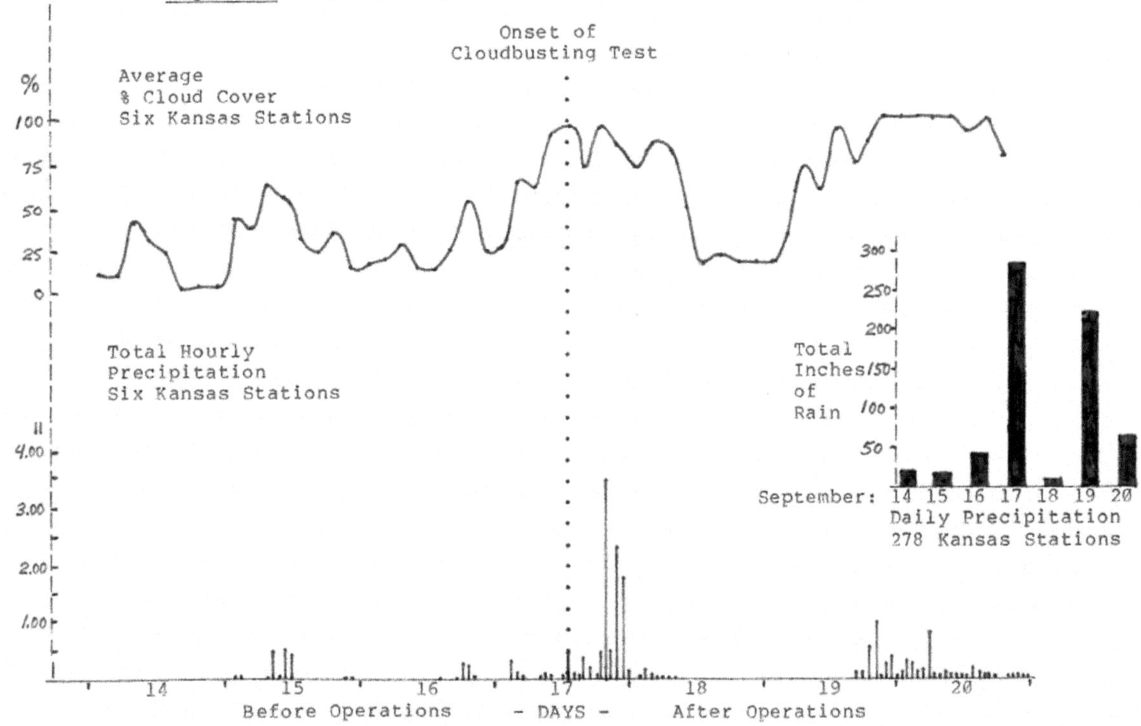

Figure 40: PRECIPITATION ENHANCEMENT OPERATION -- 24 September 1978

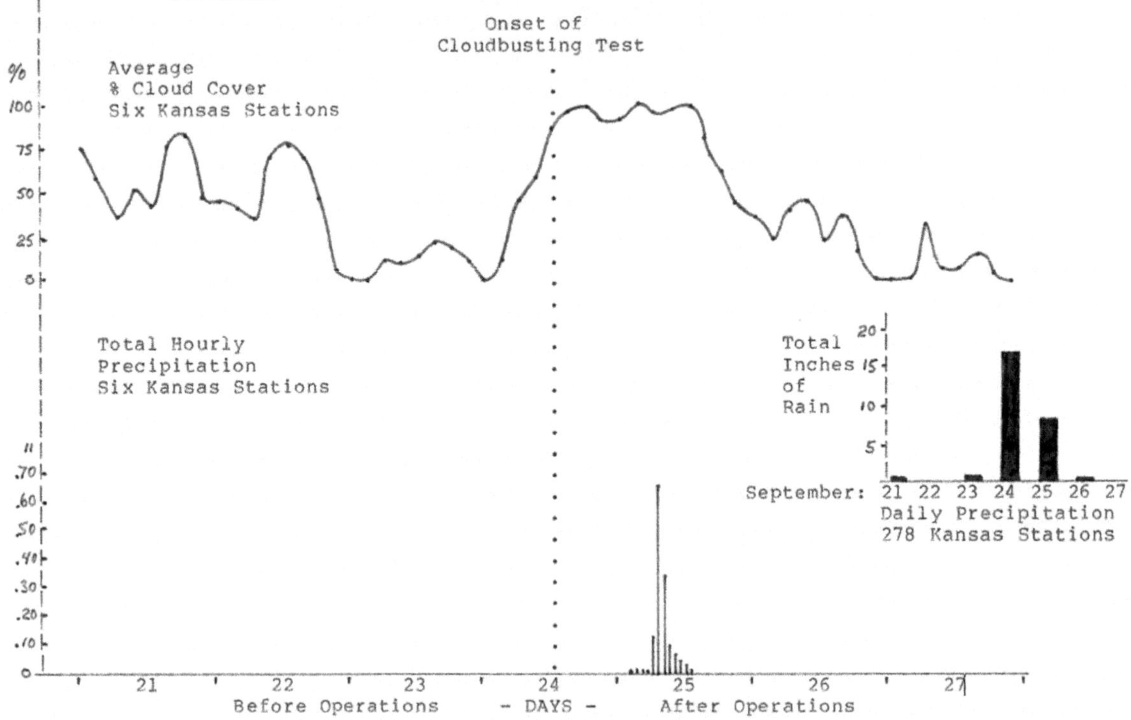

Figure 41: PRECIPITATION ENHANCEMENT OPERATION -- 1 October 1978

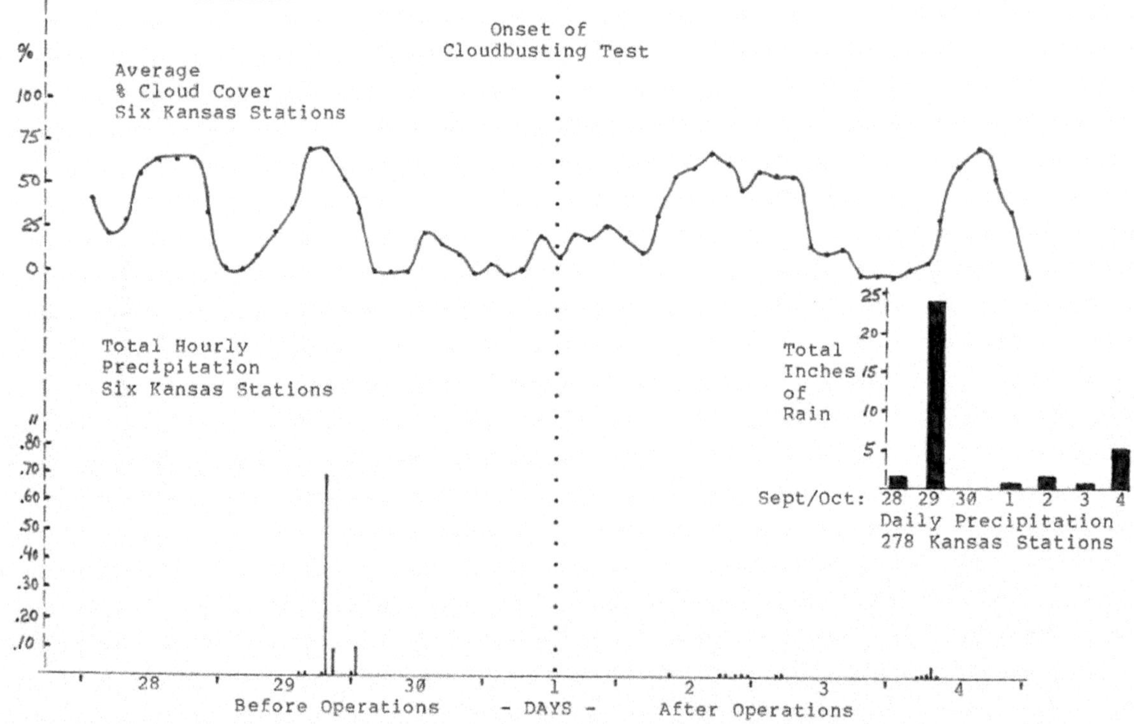

Figure 42: PRECIPITATION ENHANCEMENT OPERATION -- 8 October 1978

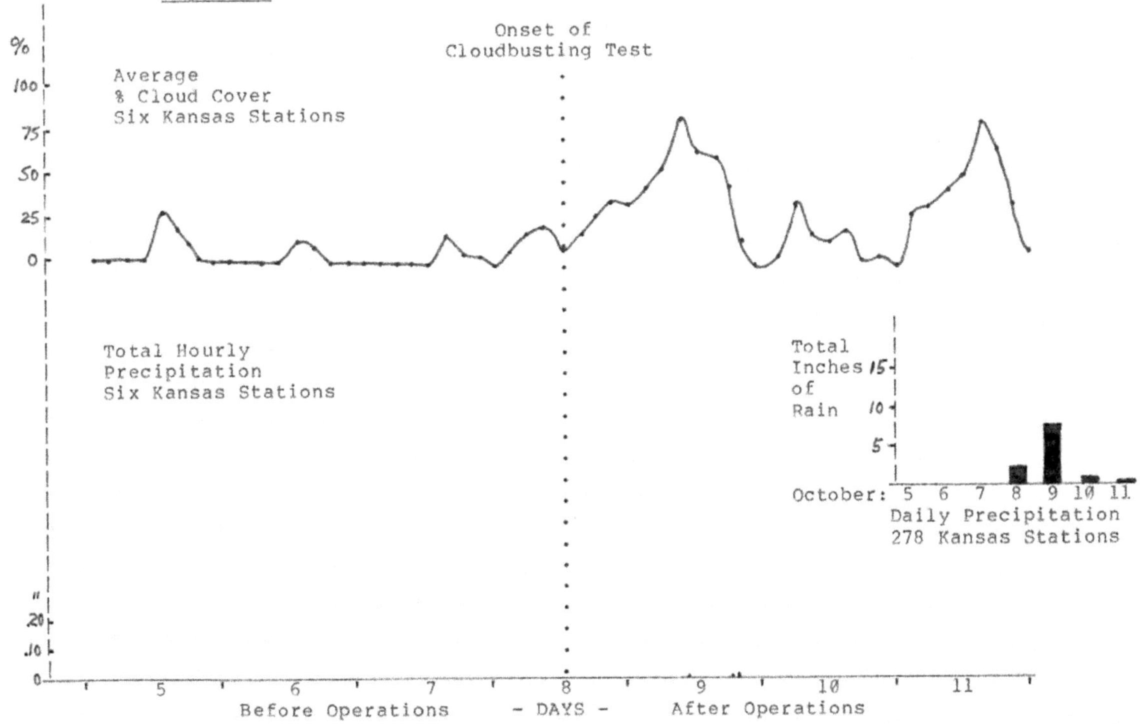

Figure 43: PRECIPITATION ENHANCEMENT OPERATION -- 15 October 1978

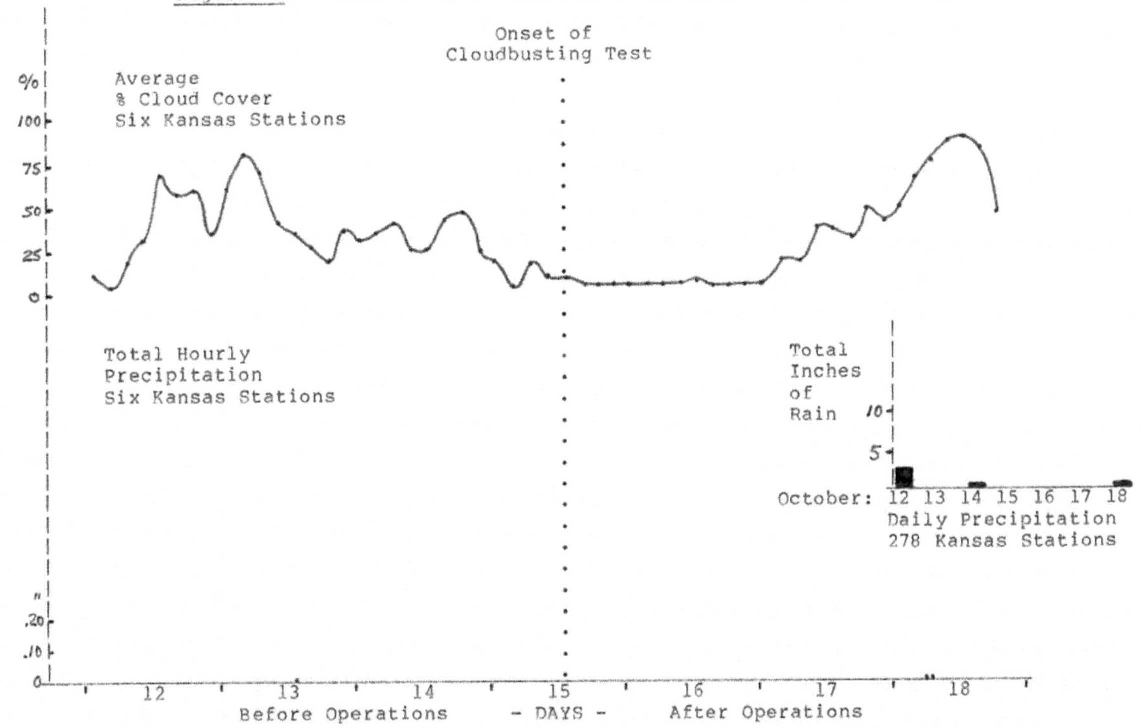

From the Figure 31 summary, we may note the following:

1. The greatest peak of daily precipitation for the summarized week, from three days before to three days after the cloudbuster test, occurs on the day of the cloudbuster operations. We recall that this daily precipitation data primarily reflects a 7 am to 7 am observational period, indicating that most of the rainfall in the peak had occurred prior to 7 am of the morning following operations.

2. Hourly precipitation data also indicates that most of the rainfall in this daily peak occurred after initiation of the cloudbuster operations, tailing off around 7 am the following morning.

3. Percent cloud cover data shows a distinct increase or "hop" in trend associating clearly with the hour of onset of cloudbuster tests. We note that for three days prior to operations a general diurnal pattern of cloud cover exists. This diurnal pattern is perturbed exactly at the time cloudbuster operations were started; it remains perturbed for the remaining three days of the observational period. This is exactly opposite of the precipitation abatement operations, where a diurnal pendulation in cloud cover was restored after commencement of operations.

Appendix IV contains the 7:00 am E.S.T. daily weather maps of North America for the morning of the day of operations, before they commenced, as well as for the morning of the day after operations. From these data we may note the change in height of the 500 mb pressure surface from 7:00 am E.S.T. on the day of operations to 7:00 am E.S.T. for the day after operations. Table III summarizes these changes below:

Table III: CHANGES IN ALTITUDE OF THE 500 mb PRESSURE SURFACE

#	Operational Date	Change in Feet
1	14 July 1977	0
2	17 July 1977	- 50
3	30 July 1977	-100
4	8 Sept 1977	-200
5	26 Aug 1978	-150
6	3 Sept 1978	0
7	10 Sept 1978	-100
8	17 Sept 1978	-100
9	24 Sept 1978	- 75
10	1 Oct 1978	-100
11	8 Oct 1978	- 25
12	15 Oct 1978	+ 75

We note that only one of these 12 operations resulted in an increase in altitude of the 500 mb pressure surface; nine of the 12 operations associate with decreases in elevation averaging - 89 ft., while two remain unchanged. (Total average for the 12 operations is - 60 ft.)

In order to analyze each cloudbuster operation separately, we can look at the data in Figures 32 - 43, at the satellite imagery for the dates of operation, in Photo Sequences 7 - 18, and at the daily weather map data in Appendix IV, asking the following questions:

A. Do satellite images show a "blossoming" effect or frontal activity in the area within 24 hours after onset of operations?

B. Does a peak in daily precipitation data occur either on the day of or on the day after operations?

C. Was this the largest peak in rainfall for the entire week?

D. Does the heaviest hourly rainfall for the week occur after the operation?

E. Does hourly rainfall increase noticeable at the time the operation began?

F. Does the greatest percent cloud cover of the week occur after operations?

G. Does percent cloud cover increase noticeable at the time the operations began?

H. Did a decrease in elevation of the 500 mb pressure surface occur, relative to the morning before operations and the following morning?

Using the above eight questions we can create a table to "rate" each operation. An operation which scores a "yes" on most of the questions is significant, while a "yes" for only one or none of the variables would be insignificant.

Table IV summarizes the answers to these questions for the 12 precipitation enhancement operations.

Table IV: ANALYSIS OF VARIABLES

Question?	OPERATION (see page 76) 1	2	3	4	5	6	7	8	9	10	11	12	Number YES per question
Do Satellite Images show response?	Y	Y	Y	Y	Y	Y	Y	Y	Y	Y	Y	N	11
Rain Peak day of or after operation?	Y	N	Y	Y	Y	Y	N	Y	Y	Y	Y	N	9
Was this the largest peak of week?	Y	N	N	N	Y	Y	N	Y	Y	N	Y	N	6
Hourly rainfall heaviest after test?	Y	N	Y	Y	Y	N	Y	Y	Y	N	Y	N	8
Noticable increase in hourly rain just after start of test?	Y	N	Y	Y	Y	N	Y	Y	N	N	N	N	6
Greatest % cloud cover after test?	Y	?	?	Y	?	Y	?	Y	Y	Y	Y	N	7
Noticable increase in % cloud cover during operation?	Y	?	Y	Y	N	Y	N	N	N	?	Y	N	5
Decrease in 500 mb pressure surface elevation after test?	N	Y	Y	Y	Y	N	Y	Y	Y	Y	Y	N	9
Total YES per operation :	7	2	6	7	6	5	4	7	6	4	7	0	

Possible points	0	1	2	3	4	5	6	7	8
Number of operations with a given point total	1	0	1	0	2	1	3	4	0

From Table IV, we note that:

1. Eight of the 12 operations (67%) show a positive response for five or more of the eight questions asked; more than half of the operations fall into the "significant" category, showing a positive response for six or seven of the eight questions.

2. Two of the 12 operations show positive response for only two or fewer of the questions. These are the two operational dates previously mentioned that did not employ a compass draw to the N.W.; only a zenith draw was used during operations.

3. Nine of the 12 operations (75%) show a peak in rainfall on the day of or the day after operations. In six of the 12 operations this rainfall was the greatest amount for the whole week.

Natural Weather Conditions

To more clearly understand the role of the cloudbuster to these interesting data, we are forced to address the following considerations:

1. Do 7:00 am E.S.T. daily weather maps made on the day of operations, before operations commenced, indicate a probability of inclement weather associated with approaching frontal systems? Looking at morning weather maps for the dates of operations, we can note the locations of advancing fronts, relative to the cloudbuster test site in N.E. Kansas. For the 12 operations, we may note five categories of frontal locations. The first category, where advancing fronts lie across Nebraska to the N.W., should be the most favorable for evening weather on the day of operations, while the last category, when no advancing fronts are over either the Rocky Mountains or the Great Plains, should be the least favorable for such weather. Table V lists the five categories of frontal locations along with their associated operational dates and their corresponding positive-response totals for variables as taken from Table IV.

Table V: LOCATIONS OF NEAREST ADVANCING FRONTS

7:00 am E.S.T. conditions	Test dates with these conditions	pos-response totals/test (Table IV)	average pos-response
Fronts across Nebraska	14 July 77	7	
	30 July 77	6	6.5
Fronts across Kansas	17 Sept 78	7	
	24 Sept 78	6	
	15 Oct 78	0	4.3
Fronts across Rockies-Wyoming	26 Aug 78	6	
	1 Oct 78	4	5.0
Fronts to North-Dakotas	17 July 77	2	
	3 Sept 78	5	
	10 Sept 78	4	3.7
No advancing Fronts	8 Sept 77	7	
	8 Oct 78	7	7.0

From Table V we note that:

A. The two operations with the lowest total positive responses to variables in Table IV occurred when fronts were either over Kansas or to the North over the Dakotas.

B. Two of the four operations with the highest positive-response totals occurred when no advancing fronts were over either the Great Plains or the Rockies.

Interestingly, daily weather maps for the day following these two operations indicate the existence of a newly-formed front over the state of Kansas (see Appendix IV).

If all the noted effects were solely due to natural weather, we would expect the operations with the lowest positive-response totals to occur when no fronts were in the area; similarly we would expect the highest point totals to occur when frontal passage was most likely for the afternoon of the test date. Such was not the case. Positive responses to variables in Table IV were just about as likely to occur when fronts were advancing from the N.W. as when no frontal boundaries existed over the Great Plains or Rockies.

2. One could set for an argument to the effect that the peaks in rainfall data in the summary, Figure 31, are due to singularly large amounts of rainfall on one or two days in the 12 operational-week study periods. We can overcome this argument by plotting all data as a percent of each weeks total, thereby giving each individual test an equal percentage weight in the summary. Figures 44 and 45 plot these percent precipitation data summarized for operations in 1977 and 1978 respectively. The 1977 data clearly indicate a peak in rainfall on the operational date. The 1978 data shows a peak in percent rainfall data on the day of and the day after operations; evidence (Figure 31) indicates that the peak in percent rainfall for days after operations is of an early morning character and is connected with rainfall which started the previous afternoon or night.

Figures 44 and 45 give indication of the possibility that there is a natural cycle in precipitation which is punctuated 180' out of phase with the peaks in precipitation associated with onset of cloudbuster operations. The 1977 data, in Figure 44, are gathered from 4 "random" operational dates and do not reveal a true cycle. But data for 1978 were systematically gathered from a sequence of eight operations performed a week apart. Is it possible that a weekly precipitation pattern similar to that of Figure 45 existed for weeks prior to our eight weeks of operations in 1978? Is it possible that a natural pattern of precipitation existed, and that we have only by fortuitous accident chosen to perform the cloudbusting operations exactly prior to a naturally occurring peak? A positive response tends to be ruled out by both the absence of a mechanism to cause such a seven-day cycle, and by analysis of similar daily precipitation data for an 8-week period before operations and for eight weeks after operations in 1978.

Figure 44: 1977 SUMMARY

4 PRECIPITATION ENHANCEMENT OPERATIONS

Inches of Precipitation as a Percent of the Weeks Total

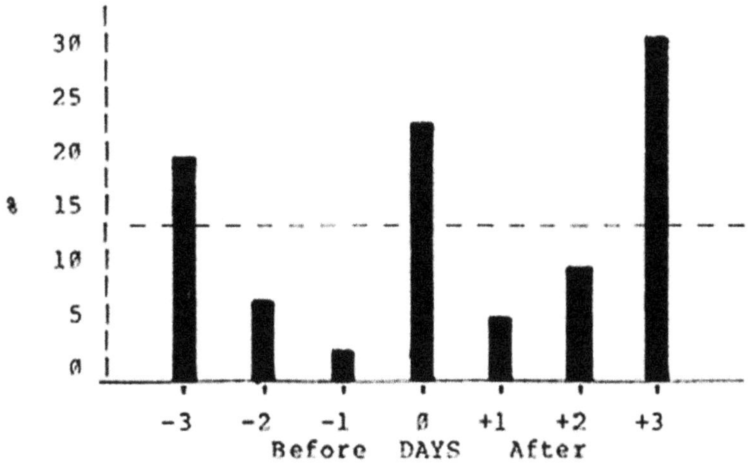

Theoretical Average - - - - - - - -

Figure 45: 1978 SUMMARY

8 PRECIPITATION ENHANCEMENT OPERATIONS

Inches of Precipitation as a Percent of the Weeks Total

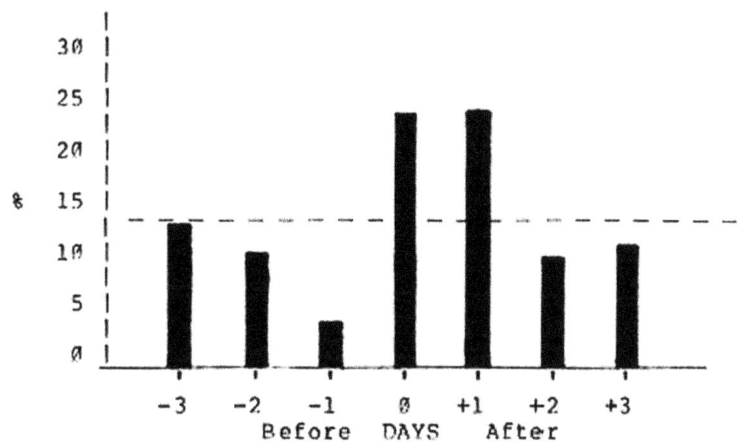

Figures 46 and 47 depict daily precipitation data, for 24 weeks in 1978, as a total quantity and as a percent of each weeks total, respectively. These data are separated into three 8-week periods occurring before, during and after the eight cloudbuster operations. If anything, we may note a dissimilarity in patterns of weekly precipitation for the three periods in question. If a natural weekly cycle in precipitation existed prior to start of the 1978 series of cloudbuster tests, it was not similar in character to that noted after tests began.

These data, as shown in Figure 46, further indicate that the greatest amount of precipitation for any day of the week in any of the three 8-week periods analyzed occurred during the test period on the day of cloudbuster operations. Similarly, Figure 47 indicates that the greatest percent precipitation for any days of the three 8-week periods occurred during the test period, for the two consecutive days on and immediately after cloudbusting operations.

From Figure 47, we see also that nearly 50% of all rainfall for the 8-week period of operations occurred on the day of and after cloudbusting operations. The theoretically expected percent precipitation for the two days would have been 28.6%.

Figure 46: SUMMARIZED DAILY PRECIPITATION TOTALS

For Three 8-Week periods in 1978

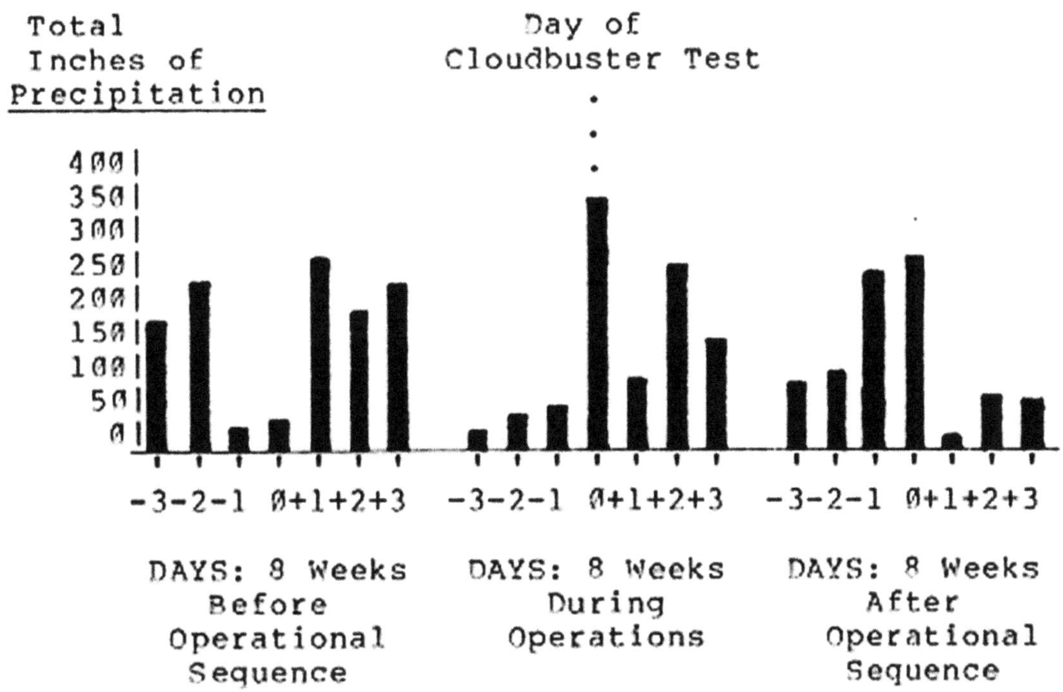

Figure 47: SUMMARIZED DAILY PRECIPITATION PERCENTAGES

For Three 8-Week periods in 1978

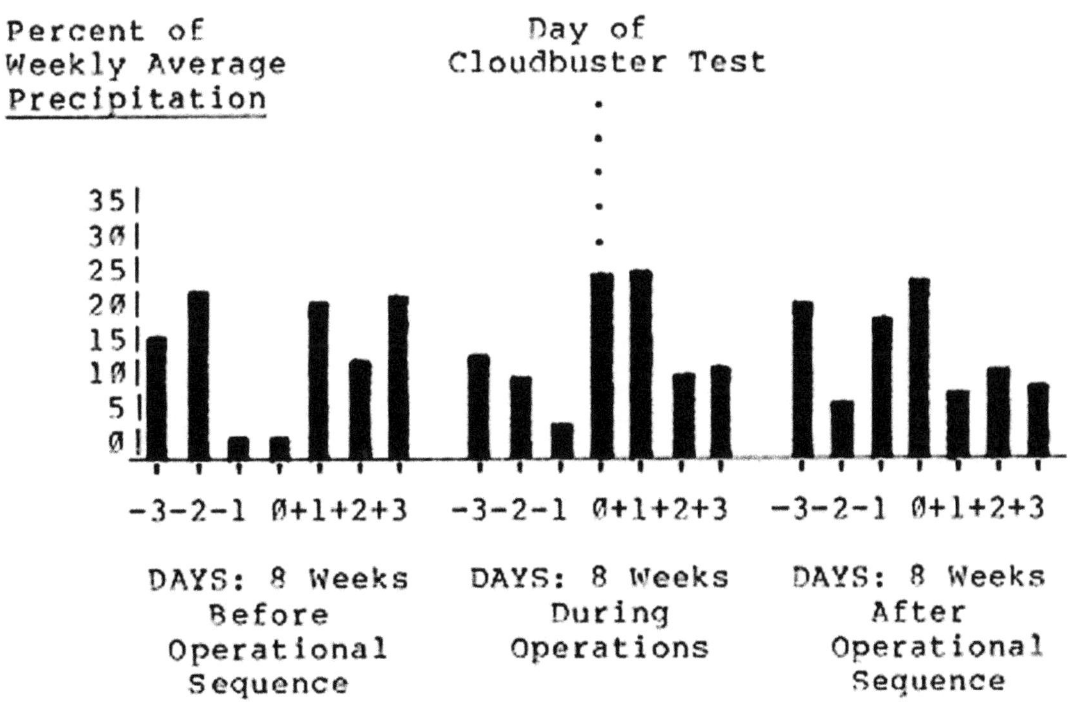

CHAPTER V

CLOUDBUSTER OPERATIONS AND SEVERE STORMS

Using Storm Data for the state of Kansas, we can see if cloudbuster tests or major operations related here generally resulted in an increase in severe storms or unusual weather (tornadoes, funnel clouds, wind storms, hail, severe thunderstorms, etc.). Occurrence of these storm events does not necessarily indicate property damage or injuries.

Figure 48 depicts the total number of severe storm events in Kansas for a week spanning the 12 precipitation enhancement operations, centered on the day of operations. Figures 49 and 50 depict similar data for three precipitation abatement operations and for 20 cloud dissipation tests, respectively. These data do not clearly indicate an increase in numbers of severe storm events associated with any of the noted types of operations. In all cases, more severe storms occurred on days before the operations commenced.

Anyone who undertakes modification of an environmental parameter as all-pervasive as weather should do so with the utmost caution and sense of responsibility. The cloudbuster operator is, in a sense, much like an atmospheric "practitioner"; what one does to the skies touches everyone else, for better or worse. While the cloudbusting technique's most beneficial direction, theoretically, is toward drought abatement and reversal of the desertification process, it would not be altogether undesirable to provide local "symptomatic" rainfall, provided that such operations do not have severe repercussions in weather events later on or at distant locations. In all cases, the ultimate goal must be toward restoration of atmospheric self-regulation, the natural state of rain-dry-rain pendulation.

It has been shown that, if the cloudbuster works at all, its affect extends over wide areas. Also, as should be apparent from the log notes (Appendix V) and as maintained in literature on cloudbusting (Appendix III), it is known that the operator's perceptual faculties play an important role in how operations are performed, and in safety of outcome. Hence, it is of prime importance that the technique, which is cheaper and more easily engaged in than cloudseeding, not become dominated by those whose perceptions are blocked: specifically by any impotent individuals who wish, in a secret manner, to play God. How to keep this technique in the hands of responsible people, in the face of an increasing acceptance of Reich's work, is a question of utmost importance.

Figure 48: 12 PRECIPITATION ENHANCEMENT TESTS

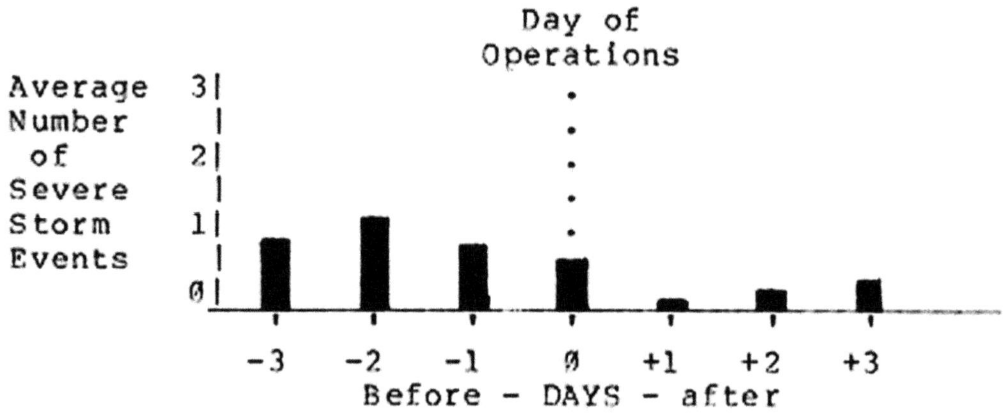

Figure 49: 3 PRECIPITATION ABATEMENT TESTS

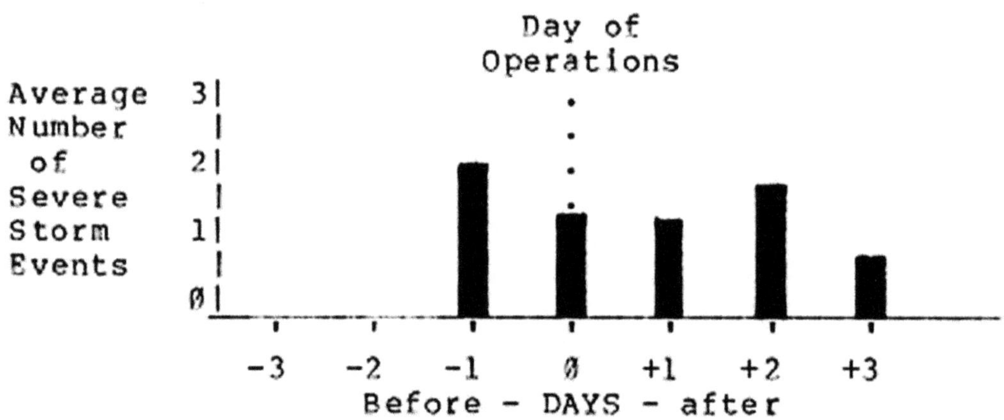

Figure 50: 20 CLOUD DISSIPATION TESTS

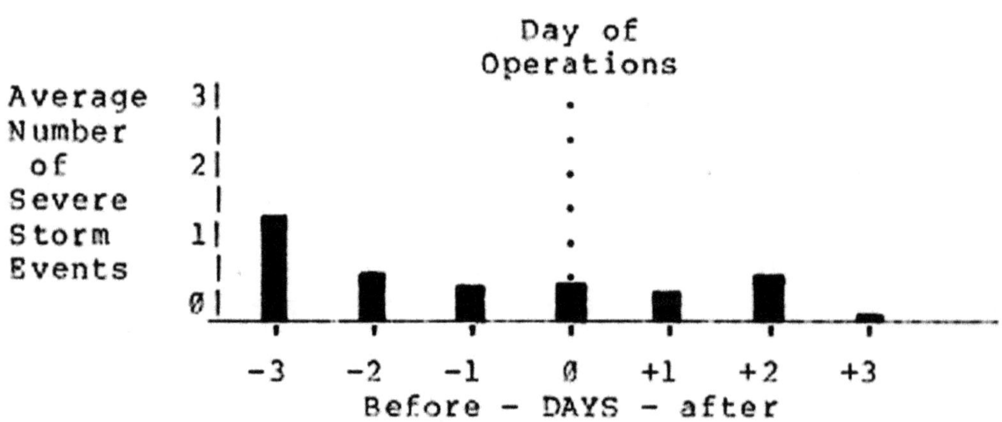

CHAPTER VI

CONCLUSIONS

When these cloudbuster tests were started in 1977, I had no firm idea of just what the outcome would be. I felt confident that some effects would be observed, based on prior experience. And certainly, more effects were observed than were objectively documented, as reported here.

Controlled cloud dissipation trials revealed interesting data which, when creatively manipulated, gave indication of increased decay rate in cumulus cloud patterns within a minute after the cloudbuster was aimed. Yet, the small number of trials leaves much to be desired. Precipitation abatement operations were also too few in number to be of much value, although a general reduction of precipitation was observed. Precipitation enhancement operations provided the most significant associations between onset of operations and summarized weather data:

1. On the average, the greatest peak of daily and hourly precipitation for the entire week encompassing test dates occurred a short time after operations were started.

2. Percent cloud cover increased and a diurnal pattern of percent cloud cover disappeared right after cloudbusting operations began. This was an exact reversal of patterns observed in precipitation abatement operations, where diurnal cycling of cloud cover was restored.

3. Barometric pressure fluctuations were recorded during one operation which associated strongly with onset and cessation of the cloudbuster operation, with in situ cloud cover, and with sensory parameters.

4. A decrease in elevation of the 500 mb pressure surface at the test site was noted for nine of the 12 tests.

We have not fully explored here the possibility that positive result in the precipitation enhancement operations was solely due to naturally occurring weather. Such is always a strong possibility when one deals with a subject as changing and dynamic as the weather. But it has been demonstrated, in the precipitation enhancement operations, that frontal location was not wholly adequate to explain all the cloudbuster associated weather, nor was a periodicity in precipitation already occurring prior to most of the related tests which could explain the observed effects and data.

Natural weather patterns could be used to explain these data only via assuming, in general, that we coincidentally decided to operate the cloudbuster a short time prior to when percent cloud cover and hourly precipitation spontaneously began an increase toward their largest values for the entire week. Regarding cumulus cloud dissipation tests, we would also have to assume that tested clouds, for some reason, were more prone to spontaneous "natural" decay just after the cloudbuster was pointed at them, as compared to control clouds observed immediately beforehand. The repetitive, non-random coincidental linking of these changes in weather patterns (cumulus behavior, percent cloud cover, hourly and daily precipitation) to the time of onset of cloudbuster operations would imply that either:

1. The cloudbuster did effect the weather, or that

2. The cloudbuster operator, who is not a weather forecaster or "seer", unconsciously knew in advance on what day of the week and at what times of day such phenomena had their greatest likelihood of occurring anyway, or that

3. The associations are purely chance events. The noted weather would have occurred exactly as it did irrespective of the cloudbuster tests.

It is possible that more than one of the above factors were at work to some extent. However, we stress the preliminary nature of this study, which should not be considered as conclusive regarding the effectiveness of the cloudbuster. Irrefutable evidence that the cloudbuster does indeed cause changes similar to those noted here would have to contain a larger number of tests in the sample, weather data gathered at the test site, reproducible laboratory modeling of the effects, plus a full theoretical integration of cloudbusting with conventional meteorological phenomena. Time-lapse movies of the effects could be most convincing, also. This preliminary study will hopefully be followed up with in situ recorded measurements of the following parameters relative to a number of cloudbuster tests:

1. Barometric pressure

2. Wind speed and direction

3. Temperature

4. Air-ion count

5. Relative Humidity and precipitation

6. Insolation: direct and diffuse

7. Earth - atmosphere electrical potential gradient

Similarly, tests should be expanded to see if a miniature cloudbuster could:

1. Have a measurable effect upon a living system, such as seed sprout length and water uptake rate, or

2. Cause unexpected changes in the following properties of water:

   Surface tension,
   resistance and conductance,
   evaporation rate,
   temperature.

The operations related here, as well as any future operations, could be also be evaluated further regarding forecasted vs observed weather patterns. Also, the study area could be widened. There are a number of approaches that could be tried to test the cloudbusting technique which were not attempted in the present study.

In closing, I feel a strong case has been made that something occurred in the atmosphere when the cloudbuster was operated during these trials. While a high degree of statistical significance has not been uniformly achieved here, nor have the most useful analytical techniques necessarily been employed, it is certain that, when the cloudbuster was operated in these trials, the atmosphere displayed a large number of repetitive coincidental changes. Natural weather conditions do not seem sufficient to explain all of the observed phenomena and weather patterns which occurred when the cloudbuster was activated. I believe, while the present study is by no means a wholly conclusive one, that the phenomena and data reported here more easily lend themselves to a positive interpretation of the cloudbuster's effect than a clearly negative one. These results clearly indicate a need for further in-depth study on the question of cloudbusting and on Reich's works and theory in general.

Although much work remains to be done, we are a long step here from the hollow legal decisions handed down against Reich's work in the 1950s. "Undeniable and conclusive" evidence for the efficacy of the cloudbuster will have to wait for future study, which inevitably must encompass and test Reich's theoretical formulations. "The Truth", Reich often said, "can only be determined in the laboratory". It is hoped that the future will allow such opportunity, particularly with regard to Reich's theoretical formulations on orgone energy.

LITERATURE CITED

Anon. (1977): "Weather Report (An Indictment) ", *Acres, U.S.A.,* November, Vol. 7, #1, Raytown, Missouri.

Bates, F.C. (1962): "Tornadoes in the Central US", *Trans. Kansas Acad. Sci.*, 65:215-246.

Blasband, R. (1969): "Orgonomic Functionalism in Problems of Atmospheric Circulation: Part I, the Normal Atmosphere", *Journal of Orgonomy,* 3:166-187, New York.

———— (1970): "Orgonomic Functionalism in Problems of Atmospheric Circulation: Part II, On Drought", 4:62-67; " ... Part III, On Desert", 4:167-182, *Journal of Orgonomy,* New York.

———— (1971 - 1979): "CORE Progress Reports", *Journal of Orgonomy*, in various issues each year.

———— (1972): "OROP Hurricane Doria", *Journal of Orgonomy*, 6:80-83, New York.

Blasband, R. and Rosenblum, F. (1972): "An Analysis of the United States Food and Drug Administration's Scientific Evidence Against Wilhelm Reich", *Journal of Orgonomy*, 6:207-231, New York.

Brier, G.W. (1955): "Seven Day Periodicities in Certain Meteorological Parameters during the Period 1899-1951", *Bull. Am. Met. Soc.*, 36:265-277.

Chalmers, J.A. (1967): *Atmospheric Electricity,* Pergamon Press, London.

Constable, T. (1971): "Rainmaking Operation Backwash", *Journal of Orgonomy,* 5:192-198, New York.

———— (1972): "Operation 'Kooler' — Conquest of a Heatwave", *Journal of Orgonomy,* 6:84-97, New York.

Eden, J. (1972): "Drought Relief in the Northwest", *Journal of Orgonomy*, 6:98-103, New York.

Eden, J. (1973): "Ufo's, Dor and Drought in the Northwest", *Journal of Orgonomy*, 7:246-253, New York.

———— (1978): "Atmospheric Murder", *Journal of Orgonomy,* 12:259-263, New York.

Greenfield, J. (1974): *Wilhelm Reich Vs. the U.S.A.*, W.W. Norton & Co., New York.

Grossmann, D. and Grossmann, W. (1955): "Wind Flow and Orgone Flow", *Cosmic Orgone Engineering*, 7:114-129, Rangeley, Maine.

Kelley, C. (1955): "Orgone Energy and Weather", *Cosmic Orgone Engineering*, 7:54-67, Rangeley, Maine.

———— (1961): *A New Method of Weather Control*, Radix Institute, Santa Monica.

Langmuir, I. (1961): *The Collected Works of Irving Langmuir, Vol.10: Atmospheric Phenomena; Vol. 11: Cloud Nucleation*, Pergamon Press, New York.

Lentner, M. (1972): *Elementary Applied Statistics*, Bogdon & Quigley, Belmont, California.

MacDonald, J.E. (1958): "The Physics of Cloud Modification", in *Advances in Geophysics* 5, Academic Press, New York.

McCullough, R. (1955): "Antibiotics, Cloudseeding and Life Energy", *Cosmic Orgone Engineering*, 7:40-46, Rangeley, Maine.

Moise, w. (1955): "OROP Drought Atlantic Coast — Summer 1954", *Cosmic Orgone Engineering*, 7:68-83, Rangeley, Maine.

Reich, W. (1942): *Discovery of the Orgone, Volume 1: Function of the Orgasm*, Orgone Institute Press, New York.

——— (1948): *Discovery of the Orgone, Volume 2: The Cancer Biopathy,* Orgone Institute Press, New York.

——— (1952): "Dor Removal and Cloud-Busting", *Orgone Energy Bulletin*, 4:171-182, Rangeley, Maine.

——— (1954): "OROP Desert", *Cosmic Orgone Engineering*, 6:1-140, Rangeley, Maine.

Reich, W. (1955a): "Expedition OROP Desert Ea, I", *Cosmic Orgone Engineering*, 7:5-19, Rangeley, Maine.

——— (1955b): "OROP Hurricane Edna - Log Notes", *Cosmic Orgone Engineering*, 7:84-92, Rangeley, Maine.

——— (1957): *Contact With Space: Oranur Second Report (OROP Desert Ea),* Core pilot Press, New York.

Silvert, M. (1955): "Report of OROP Desert Ea Survey of Tucson Area", *Cosmic Orgone Engineering*, 7:200-202, Rangeley, Maine.

U.S. Department Of Commerce (1972): *Summary Report: Weather Modification, FY 1972*, Government Printing Office.

——————————— (1976): *Weather Modification Activities for Calendar Year 1975,* National Oceanic and Atmospheric Administration, Government Printing Office; cf ibid, 1972, 1973, 1974.

Wolfe, T. (1948): *Emotional Plague Versus Orgone Biophysics*, Orgone Institute Press, New York.

APPENDIX I

EVIDENCE FOR THE EXISTENCE OF A PRINCIPLE OF ATMOSPHERIC CONTINUITY

Table of Contents:

I. INTRODUCTION

The concept of a ubiquitous energy continuum in the atmosphere and in space is not a new idea. It has been embodied in many world religions, primarily of eastern origin for thousands of years, and evolved in Western thought to the form of a luminiferous cosmic ether (or aether). However, with the rise of scientific specialization and atomic thinking, the natural world was more and more viewed as an assembly of disparate parts, functioning in a mechanical manner. Matter was viewed as a conglomerate of discrete particles with "nothing" existing in between.

The experiments of Michelson and Morley (1887) were thought to have done away with the notion of a medium in space upon which action-at-a-distance influences were passed. Relativity and quantum electrodynamics flourished in a theoretical atmosphere devoid of any interconnecting principle which would oppose the notion of "empty intermolecular space" or the inviolability of the laws of inertia.

However, much scientific observation has consistently cropped up which will not fit comfortably into a theoretical framework of "empty" space. To date, these observations are often viewed as anomalies, or, at best, mere curiosities of science. But the pile of "anomalous" observations is quite large now, and harder to ignore.

Astronomy has theoretically languished while space scientists discovered flux and flow of energy and mass in interplanetary space, once thought to be near sterile. Because of this, astronomy textbooks written more than 10 years ago are now quite useless. Biology and meteorology have similarly failed to incorporate into their theoretical framework the interdisciplinary discovery of many diverse cycles, seemingly orchestrated by solar-planetary configurations. It turns out that, contrary to thermodynamic law, more harmony and resonant rhythm exists in the universe than the expected cacophony; chaos is the exception rather than the rule.

In the present appendix we explore evidences which support the idea of a universal medium which stretches down to us from the heavens, providing a contiguity between planets, atmosphere, earth crust and organisms. We will build a case for such a medium partly by inference, and partly by direct evidence, mentioning that time was not sufficient for fully organizing the material around a central theme.

The "medium" we discuss here is functionally identical with the principle of atmospheric continuity assumed to exist for the body of the present thesis. I view the present appendix also as the preliminary stage of a more in-depth undertaking and beg the reader's pardon for failure to theoretically integrate the material at this time. We will start with considerations of the astrophysical sphere, and eventually narrow down to atmospheric and biological phenomena.

II. THE LUMINIFEROUS COSMIC ETHER AND RELATIVITY

The notion of a substance finer than air, upon which the impulses of physical phenomena were transmitted through a distance of space, dates back to Aristotle and perhaps earlier. The luminiferous cosmic ether was postulated by Newton as a medium through which action-at-a-distance phenomena could occur. The concept was thoroughly incorporated into physical science of the late 1800s as a mechanism to explain the wave action of light. However, the ether theories were shortly displaced, around 1920, by Einstein's work, which needed no cosmic ether to explain the same phenomena. Einstein's relativity and Planck's quantum mechanics, which grew, in part, from Maxwell's equation, have currently supplanted any notion of ether in the minds of the majority of scientists in the 1970s. However, as will be pointed out here, and as stated by Nobel laureate Dirac (1951), and by DeBroglie (1959) and Dudley (1976), there is much reason to resurrect the notion of ether. Indeed, many modern atomic theories require an energy-rich subquantic medium as part of their theoretical framework.

If we view the language and concepts of modern physics and cosmology we are struck by an array of terms giving a view of the universe that might well have originated during a psychotic nightmare, or at least in a science-fiction novel: "strangeness" quantum numbers, "antimatter", fluids so volatile they can "pass through a nonexistent hole", "forbidden radiation", "charmed particles", "absolute ' elsewhere", "quarks", "black holes" and "white" holes, etc. These terms are certainly no less exotic than Reich's "orgone", or any principle of continuity. Atomic science, instead of bringing order into our perception of a seemingly chaotic universe (one of its supposed goals!), has further complicated the picture; the closer we step towards a hopeful focus, the more blurred things appear. It has been argued that an attempt to simplify any explanation of nature originates from merely a different philosophy of science, rather than from any real-world existence of simplicity. But we shall avoid philosophical argument here. Perhaps the complicated, particle-physics-relativistic view really best explains the deepest level of functioning in the natural world; perhaps it does not, and simpler explanation for seemingly diverse phenomena exists. Let us take a quick look at physical evidence which lends support to the existence of an ether in space which holds the universe and molecular matter in stable, lawful order.

Argument For a Cosmological Ether and Ether-Drift

Sir Isaac Newton postulated a static ether which could act as a frame of reference for the motion of all objects. With this frame of reference he worked out his laws which have been a cornerstone of modern scientific thought. Why Newton needed a static, non-moving cosmic ether is an interesting question, approached by deGrazia (1966) and Kubrin (1972), which has to do with Newton's theology. We shall pass over theological considerations due to space limitations. However, we must note that, had Newton been able to postulate a dynamic ether in movement with the heavenly bodies rather than acting to them as air does to a bullet, much of the history of science might have been different.

Much has been learned about the physical universe since Newton's time, particularly about the phenomena of light. As light exhibited wave properties, similar to that of sound or water waves, the ether took on a real significance in explaining just what the light waves were waving in. Further, the implications for cosmology, in explaining how light and heat came to us from the sun and from stars was of enormous importance. Hence, a great amount of energy was exerted toward finding and proving the existence of the hypothetical ether, and in isolating its properties. To this end, Michelson (1882) experimentally occupied himself. He was later joined by others in a series of what came to be known as the ether-drift experiments.

The ether-drift experiments were an attempt to show that the earth speeded through a medium in which light waves were transmitted. Calculations stemming from measurements of stellar aberration led Michelson to believe that the cosmic ether could be detected, and the earth's relative motion to it determined. Michelson built a device called the interferometer, which could accurately measure slight differences in the speed of light beams traveling in different directions. His calculations led him to believe a shifted pattern of "interference fringes" would be detected with the device, proving the existence of the ether. A pattern of interference fringes shifted 4/100ths of a fringe would vindicate Newton and the "static" ether. Actually, he detected shifts of only 2/100ths, and concluded only that a "static" ether did not exist (Michelson 1881).

However, the questions involved here were so great that controversy dominated the scientific atmosphere. Michelson and Morley (1887) again duplicated the experiment, this time with a more sensitive apparatus. Shifts were expected to be 4/10ths, but a shift of only 1/100th was observed. This tended to lay the question of a fixed or static ether to rest, and also to leave theoreticians in a confused muddle. The ether had been used to explain a number of phenomena and now the notion of it was under fire, even though Michelson and Morley felt they had only disproved the existence of a static ether. The small fringe shifts that were detected could be retained into an ether theorized as being dynamic in nature, moving along with the earth close to the earth's rotational velocity.

It is interesting that history generally records a "negative" result for the Michelson-Morley 1887 experiment, even though fringe shifts of some significance were detected. The 1887 experiment, which gave a slight positive result, is too often reported as having had a null or zero result, which is simply not true. The experiment was and still is widely misinterpreted, being hailed erroneously as the experiment which proved the nonexistence of the ether and which gave impetus to widespread acceptance of relativity. Relativity is a mutually exclusive theory regarding an ether in space or a variable speed of light (which an ether might cause) . Checking the record, Michelson and Morley performed the ether-drift experiment of 1887 on only three days in July, at noon and evening for one hour each, a grand total of six hours worth of measurement They stated a desire to repeat the experiment later so as to record periodic seasonal differences in the slight positive result obtained, but never followed through with it. Further, their work was executed in the basement of a massive building. As we shall discuss, these factors of altitude, shielding and seasonality (as well as misinterpreting a slight positive result as a "null" result) were most crucial for the history of science.

Swenson's study (1972) of the cosmic ether question depicts a scientific world in early 1900 as badly needing a theory to explain the mechanism of phenomena once covered well by the idea of a subtle medium or ether. New phenomena of light, revealing a particulate nature, were also emerging, needing explanation. Even the discovery of Hertzian radio waves could not be considered as a proof of existence of the ether; this work grew from Maxwell's equations and hence needed no ether to explain it. The stage was set for Lorentz and, later, Einstein (1905). Michelson lost interest in the ether question, busying himself in pursuit of practical applications for his interferometer, such as measurement of the speed of light.[1]

Lodge approached the question of a dynamic ether in 1881, experimentally testing the idea that a massive round body in motion could alter the speed of the light passing nearby, due to forced motion of the ether entrained at the surface of the body. He interpreted his results as being negative, but our present perspective shows he failed to adequately proportion the scale of his experimental apparatus to that which nature exhibits. Morley and Miller (1904) carried out further tests designed to explore the questions raised by Lorentz (1889) regarding contraction of mass in the direction of the ether drift. These experiments of Morley and Miller were carried out in the same basement used by Michelson and Morley in 1887. Again, a slight positive result was obtained (113/1000 expected, 15/1000 obtained). Trouton and Nobel (1902), Rayleigh (1904) and Brace (1904) similarly attempted to duplicate the experiments but reported a negative result.

---

1. Even Michelson had "given up" on the question of the non-constancy of light, in spite of his own experimental results. (2010 correction: Michelson died convinced of the existence of the ether. See DeMeo publications from 2002-2004 on p.xviii)

However, the question was not settled in everyone's mind, as the notion of a dynamic ether still had to be dealt with, as did the existence of a slight positive result obtained by nearly everyone. Morley and Miller again tackled the question by experimenting in a glass room high up on Mount Washington. They felt this would allow better transmission of the ether in a location where the laminar flow of ether near the earth would be minimized. They overcame the objections which had previously been raised regarding massive building materials and altitude. By 1905, Morley and Miller announced a slightly larger ether-drift measurement, smaller than predicted by Newtonian mechanics, yet, as Miller put it, "a very definite positive effect".

Einstein's work on relativity also came out in 1905, requiring, as one of its several unproven assumptions, constancy in the speed of light. Morley soon forgot the ether but Miller continued to pursue it. Meanwhile, Sagnac (1913) published two reports of a rotational ether-drift experiment giving positive results, as did Zeeman (1915), Righi (1919) and later Miller (1921). Working on Mt. Wilson, Miller reported "strong definite indications four times as large as before". By 1923, Miller had perfected his temperature-controlled interferometer technique and was able to detect an ether drift of 20% - 30% of that expected for a static ether. Further, his data, which was accumulating over the years, showed inexplicable but definite periodicities ranging from as small as 25 seconds to that of a yearly cycle.

Miller also calculated from his ether-drift data the net movement of the Earth relative to the cosmic ether, finding it to be along a line which stretched from a point in the constellation Hercules. This is in agreement with modern astronomical observations of the earth-sun net movement relative to our own local cluster of stars.[2] Miller further reinterpreted the result of the Michelson-Morley 1887 experiment, finding the slight positive results to agree with his own directional calculations.

In 1925, Michelson again experimented with a new interferometer technique and obtained questionable results. Miller, who by now was most familiar with the effects of building materials and temperature on the experiment, was critical of Michelson's technique, which was carried out in the metal dome of Mt. Wilson observatories. The fact that metals or concrete should have an effect upon such experiments was inexplicable in 1925. Today, with the works of Reich (1948), Bortels (1951, 1956), and Piccardi (1962), such entrainment phenomena finds empirical support.

---

2. The net velocity of Earth relative to ubiquitous cosmic radiation has also been plotted (Conklin 1969).

To continue, others tried the ether-drift experiment or a variation of it. Piccard and Stahel (1926) reported a negative result. However, Illingworth (1927), Michelson (1927), and Kennedy (1935) all found positive result, yet still smaller than the static ether prediction.

Again, Swenson paints a picture of a science newly intrigued with relativity and quantum mechanics, and paying little attention to continuing ether-drift experiments. With the politics of science in full swing, ether critics now compared further work on the question to the searches for perpetual motion. Such charges, which effectively isolate a researcher from funding and publication, killed further serious work on the subject. Michelson died in 1931, with the erroneous reputation for having "proved" the nonexistence of the ether. Miller died in 1941.

It is interesting how the Michelson-Morley experiments of 18 87, which encompassed a few hours of data collection in a single week, are considered as "proof" for constancy of light speed, and for the nonexistence of the ether. Meanwhile, Miller's work, which stretched from 1905 to 1926, encompassed over 200,000 separate measurements giving significant positive results, has been virtually ignored or forgotten. Miller (1933) has given a major summary of the ether-drift experiments and his own work, arguing that interferometers used by his critics were enclosed in heavy metal casings or were in basements below the ground and in massive buildings. Again, Miller's interferometer used glass or cardboard panels and was set up at high altitude within a 360 degree glass-windowed shelter.

In 1938, Stilwell and Ives reported a positive transverse Doppler effect, but were met with nothing but philosophical argument. In 1951, Nobel laureate Dirac published a theoretical paper on the necessity for an ether to explain new phenomena in quantum electrodynamics. This induced a host of experimenters to try again the interferometer tests. Ditchburn and Heavens (1952), and Essen (1955), using radio waves instead of light, reported null results, as did Townes, et al (1963) using laser light. Shankland (1954) reinterpreted Miller's data of 1921-1926 and explained it away as a temperature effect. I have not checked out this last group of papers to see what the experimental setup was like; Shankland's statistical and temperature-effect arguments seem quite empty considering the gap in time — Miller appears to cover the question of temperature quite adequately in his 1933 paper.

DeBroglie (1960) introduced the concept of the "neutrino sea" as a substitute for the ether. At the time the neutrino had not yet been detected. DeBroglie may have been begging the question merely by changing the terminology; physicists tended to view the ether with a jaundiced eye. (see section on Neutrino Sea) Table I summarizes these various research efforts with respect to both a static and a dynamic ether.

Table I: Ether-Drift Experiments and Results

Date	Experimenter	Notes	+/- Evidence for Aether: Static?	Dynamic?
1882	Michelson	50% of that for static aether	-	+
1887	Michelson, Morley	2½% of that for static aether	-	+
1904	Morley, Miller	1.3% that for static aether	-	+
1902	Trouton, Nobel		-	?
1904	Rayleigh		-	?
1904	Brace		-	?
1905	Morley, Miller	larger than 1904	-	+
1913	Sagnac		?	+
1915	Zeeman		?	+
1919	Righi		?	+
1921	Miller		?	+
1923	Miller	20-30% of that for static aether	-	+
1925	Michelson	Metal dome, Mt. Wilson Obs.	-	?
1926	Kennedy		?	+
1926	Piccard, Stahl		-	?
1927	Illingsworth		?	+
1927	Michelson		-	+
1933	Miller	Summary, 1926	-	+
1938	Stilwell, Ives	doppler effect	-	+
1952	Ditchburn, Heavens	Radiowaves	-	?
1955	Essen	Radiowaves	-	?
1963	Townes, et al	Laser light	-	?

[NOTE ADDED 2010: The above table was compiled in 1978, and is very much out of date. While there is nothing I would remove from it, there are many new experiments on the ether-drift question, undertaken with positive results by around a half-dozen professional scientists and engineers. Notable among them is Prof. Yuri Galaev of the Ukraine Radiophysics Institute, who has confirmed the Miller result "down to the details". I have cited these in my own papers on the question of cosmic ether-drift, as published since c.2000. A comprehensive listing with citations and some PDF downloads are found at this webpage:
www.orgonelab.org/energyinspace.htm]

The above table would suggest that a static ether certainly does not exist, but that a dynamic ether does exist. At the very least, one has to admit that the possibility for existence of a dynamic ether is still an open question.

In retrospect, to my knowledge, only the work of Miller took fully into account the possible effects of altitude and material surroundings. Also, we are left with the puzzlement of slight positive result manifest in many of the experiments. If Michelson himself, a proponent of the ether theory, did not take care enough to consider building materials and altitude, how careful would his critics have been? This question remains unanswered.

Neutrino Sea?

The neutrino is a nearly massless and chargeless subatomic particle intimately associated with radioactive decay. So many of them are naturally formed it boggles the imagination. According to Asimov (1966), the sun theoretically produces 1.8 x 1038 neutrinos each second. Of these, the earth intercepts 8 x 10 28 neutrinos per cm2 per second. Similarly, the earth gives off a cousin particle, the antineutrino, at a rate of 1.75 x 1026 per second. The average human body is about 70,000 cubic cm volume, and therefore receives about 1750 antineutrinos and 2,400,000,000,000 neutrinos each second, theoretically only from earth and sun, respectively. A large nuclear reactor also puts out antineutrinos, at a rate of about 1018 per second. In order for one of these "particles" to theoretically interact with matter, it would have to pass through a hundred million times 140,000,000 miles of solid matter, or about 3500 light-years thickness of solid lead. Dudley (1976) raises the interesting question: What happened to all the neutrinos and antineutrinos produced since the beginning of time by all the different stars and planets? Due to their highly penetrating properties, most of them should still be floating or zapping around, and in incredibly abundant numbers.

Physical search for the neutrino entailed using complex sensors placed deep inside the earth. After nine months, seven reactions occurred inside the sensors and physical theory was hailed as triumphant. However, this triumph was short-lived. When the neutrino sensors were turned toward the sun, they failed to detect the expected number of solar-produced neutrinos. At present, either the existing theory of neutrinos or that of fusion as the source of the sun's energy must fall away.[3]

The neutrino was originally postulated as a bookkeeping particle to explain discrepancies in energy-mass conversions during radioactive decay. Theory simply did not (and still does not!) fit with observed phenomena. Certainly, the

---

3. "Something Wrong Under the Sun", *Science News*, 17 February 1979, p.103.

physicists researching such particles are dealing with something which is diaphanously massless and chargeless, but a case can be made for other explanations than the contradictory "particle-waves". As related previously, "neutrino" and "neutrino sea" are terms ultimately generated to substitute for the older "ether". Both the old ether theories and the newer neutrino theories link radioactive decay with *some kind of massless, penetrating force which is available for interaction with matter and other forms of energy at virtually any point or location in the universe.*

Redshift Controversy

If a medium exists throughout nature, it is likely to extend its ubiquitous property out to the stars, thereby effecting the nature of light coming to us from the far reaches of the heavens. It has been discussed how a medium for the wave theory of light was done away with prematurely, and later substituted with a more complicated "particulate" universal medium. When the first galaxies were discovered and spectral analysis of them obtained, it was noted that galaxies showed a distinct shifting of spectral bands toward the red end of the spectrum: hence the term redshift. One possible cause was given by the Doppler effect: if the galaxies were receding from us then the waves would appear slightly longer, just like the depression in sound pitch heard as a train passes by. And, since light was given a relativistic, quantum-mechanical explanation for its propagation mechanism, the receding of galaxies was the only viable explanation. Hence the idea of an expanding universe and the Big Bang theory, quasars and black holes.

However, if an interstellar medium exists, which would be to light as water is similarly to water waves, or as air is to sound waves, then another explanation could be found for red-shifting which would not wholly eliminate Doppler effects from the picture. It is well known that waves in a medium attenuate over distance due to friction and absorption of energy from the wave by the medium itself. Drop a pebble in a pond and the waves generated will gradually lengthen as they fan out from the source. The same is true for sound waves. If an interstellar medium exists, light coming to us from the stars would be redshifted (lengthened) due to the nature of the medium, which would absorb energy from the electromagnetic light wave (Rosenblum 1970).

The notion of "empty" space has come under attack from many directions. It is now well known that radio astronomers have detected the existence of methane, methyl formate, methanol, acetonitrile, methylamine, methyl acetate, etc., etc.; even the existence of interstellar cellulose has been postulated (Fox & Jennings 1977).

Arp, of the Hale Observatories, has collected instances of anomalous groupings of astronomical bodies with vastly different redshifts (Arp, et aI, 1973). High redshift objects appear in front of low redshift objects, the nearer one partially absorbing light passing nearby, even though its redshift indicates a juxtaposition of distances. Similarly, galaxies with vastly different redshifts appear connected by luminous gas, often in clusters, and sometimes appearing with quasars.

It also seems that different areas of space possess different red-shift-distance relationships. Light from a distant object seems redshifted improportionately if it travels through a cluster of galaxies. Further, many interstellar objects, such as whole galaxies, based on empty-space, constant light-speed distance calculations, seem to be expanding or splitting up faster than the speed of light (Vigier, et aI, 1974; Cohen 1977). Space scientists have provided evidence over the years to show, in stark contrast with notions of "empty space", that there is as much matter dispersed in between the stars and planets as there is contained within them.

Argument Against Relativity

Modern physics and chemistry have set up a framework dictating exactly how interrelationships occur between atoms, molecules, electromagnetic fields and other action-at-a-distance phenomena. If sufficient doubt can be cast on this scaffolding erected by theorists, then we have a legitimate case for reinterpreting existing data in another way. We are not questioning empirical data and factual observations, merely the theories generated to explain these facts.

The theory of relativity has captured the minds of most modern physicists, yet it is founded on certain assumptions, some a priori, made in the years 1900 - 1925. Our present knowledge of the universe has been extended mostly by space scientists and not by theorists. Data now exists which casts significant question on certain underlying assumptions of relativity, some of which are:

1. Riemann Geometry — parallel lines converge; sum of angels in a triangle are greater than 180 degrees. This postulate of Riemann was made in an 1854 lecture, without any evidence to support it. In the real world, if someone suggested this was the way things were, we might rightfully suggest he see an optometrist. Einstein (1916) incorporated this geometry into his theories on curved space.

2. Constancy of the velocity of light positive ether drift experiments invalidating this assumption have already been given here. Also see the section on the Redshift Controversy.

3. Two frames of reference in the common "traveling twin" examples used so often to prove relativistic dilation of time, two frames of reference are used, one for each "twin" or "clock". However, as Dudley (1959) points out:

"... when two frames of reference are used to describe two particles, in the discussion there must be a human mind on one or the other of the particles, or traveling at the same velocity as one of the particles. The particles are referred one to another not through the origins of the two systems of coordinates, but through a human mind ... man has become part of the experiment, not an observer, and by the varying possible mental condition of the thinker all manner of relationships can be 'proved'. We are reminded here of the German mathematicians of the 1800s who were converted to spiritualism because an English medium was able to produce effects in seances which were provable by n-dimensional geometry."

4. Radioactive Decay Rate Constancy — much of modern physics and chemistry is based on the assumption that radioactive decay rates are constant. Hence, when radioactive materials are accelerated to near the "velocity" of light and display change in decay rate, it is cited as a proof of relativity. Yet, substantial evidence exists to show that radionuclides can change in decay rate without such high-speed gyrations, and in real-time. Experiments have shown that events within an atom's nucleus, long thought to be independent of influence from without, can be modified by atomic changes of state (liquid-gas-solid) or chemical partnership. Changes of decay rate in a variety of elements have been reviewed by Emery (1972). Anderson and Spangler (1973) have similarly shown that decay events are interrelated and not wholly independent one from another, indicating a continuum effect. I have not had the time for full investigation of this interesting area, particularly for the works on the topic by Kurzner (1924), Curtiss (1932) and Berkson (1966). Reich (1951) also produced evidence showing that the decay-rate properties of radioactive substances could be changed.

5. Inertial Constancy — Newton's first law of inertia, that an object at rest remains at rest unless acted on by an outside force, leaves us with an ultimate question, of *where all the movement of the universe originated from.* The whole of experimentation from Newton's time on, especially with regard to subatomic physics and cosmology, can be viewed as a pursuit to finally answer this one question. Of course, much of physical theory seems secure without pondering such a question, but theories in modern physical science appear as dominoes, leaned against each other in a long line, ultimately resting on the single domino of inertia. Hence, any experimental evidence that can show the existence of a spontaneous impulse in nature, supporting the notion that nature moves without the need for a force to make it move, would solve a great many questions. Radioactivity, "random" mutations in gene theory, motivation in living systems, and weather dynamics all possess an element of spontaneity. We have covered evidence indicating an unknown element existent in radioactive processes. Regarding inertia, DePalma (1974) has shown that a spinning metal sphere will have different inertial properties from a non-spinning metal sphere. Merritt (1974) has similarly discussed non-Newtonian behaviors in counter-rotating gyroscopic pendulums. Evidence of these gentlemen further indicates

that a field effect extends outward from a rotating body which can affect inertial characteristics of nearby objects. A spinning ball bearing will rise higher and then fall faster than a non-spinning one when subjected to the same upward force within a gravitational field. Similarly, a tuning-fork-operated Acutron watch will run slower when placed near a massive, magnetically shielded gyroscope, as compared to an identical timepiece monitored elsewhere as a control. Kozyrev (1968) has similarly demonstrated non-Newtonian behavior in gyroscopes and pendulums, as have Reich (1957), Brown (1960), Saxl (1964), and Baker (1977).

III. RHYTHMS AND CYCLES

Most phenomena of the natural world undergo slight or major fluctuations of intensity. Rare indeed is a component of the natural world which can be proved entirely static. Such fluctuations may be periodic, waxing and waning in an obvious manner, or may possess a seemingly random fluctuation, which later may resolve itself into the summation of several overlapping cycles. Dewey (1970, 1971) has collected a host of such factors in many different fields of science. Many diverse phenomena can be grouped together on the basis of their periodicity. Further, most variables with a given frequency will wax and wane at the same time. The simultaneous upswing and downswing of diverse, seemingly unrelated variables suggests a common orchestrating element.

For instance, cycles of sunspots coincide with cycles of animal abundance, good wine years, rainfall patterns, etc. (to be discussed shortly). All of them possess an identical periodicity and phase. In this example, we intuitively see the dominant factor, the sunspots, as having something to do with all the Earth-based cycles of similar periodicity. Weather is obviously a major intervening factor regarding many of these interrelationships.

However, we are not allowed to rest entirely with the thought that all such cycles are solar-orchestrated. Sometimes the earth-based system peaks at an earlier time than do the solar events themselves, indicating the effect may ultimately come from outside the solar system, for both terrestrial cycle and sunspot cycle.[4] The failure to detect significant neutrino flux from the sun, and the solar-wind modulation of cosmic ray flux also hint in this direction.

Regarding terrestrial cycles, many phenomena with the same period and phase present a puzzlement by not showing any single factor that might be construed

4. Juergens (1976) discusses evidence, supported by non-empty-space astrophysical evidence, to show that the sun gets its energy from outside the solar system and not from internal fusion. Also see Dewey (1977).

as the common orchestrating element. Cycle lengths in such cases may not be traceable to sunspots, or, for that manner, to any major geophysical or extraterrestrial "zeitgeber" we know of. Of course, much work remains to be done on the matter.

It is understood that critics of cycle research feel that study of such phenomena is an exercise in coincidental statistics and is therefore meaningless. I must reject this outlook, as many once-mysterious cycles have, with time and investigation, revealed themselves locked into a distinct cosmic pattern with a milieu of other interrelated cycles. In most cases, the coinciding extraterrestrial triggers are delineated, even if the mechanisms by which such effects occur are still not understood. The literature for cycle research is too comprehensive to adequately cover here; instead we will deal only with those items which imply a natural connectedness between cosmic and atmospheric space, and a linking with living systems.

Interplanetary Influences –
One Heavenly Body Influences Another

Since we have started our scenario with an overview of questions on the nature of space, we will continue by exploring observational evidences for an affect of one heavenly body upon another. The mechanisms are not understood here, except insofar as they cannot be wholly explained by conventional theory.

The sun is orbited by several planets held in place by what we call gravity. However, having given the attraction of one mass for another a name, gravity, we understand it no better. Neither does deciphering the interactions of mass with mathematics help to solve the basic question of just what gravity is and how it works.[5] We know planetary bodies are "gravitationally" interacting with the sun, and that electromagnetic and magnetic affects participate in such interactions also. We are most familiar with the ideas relating the attraction of planets *by* the sun. However, planetary bodies can also have an effect *upon* the sun, some of which are explicable, and some not, through the medium of "gravitational" interaction.

As early as 1907, sunspot groups were suspected of being influenced by the earth's position due to regular discrepancies between the numbers of spots

---

5. Assuming one does not go for the relativistic interpretation of gravitation as a curvature of space-time. Relativity, which grew in reaction to the nonexistence of Newton's theologically-necessary static ether, places gravity in an intangible relationship to the observer (yet we know gravity is quite tangible! We feel it!). However, an ether which dynamically caused gravitation would conversely be both tangible and modifiable.

appearing with the sun's rotation and the fewer number which disappeared after passage over the solar disc. These solar phenomena prompted similar searches for effects of other planets. Increases in sunspot number were noted particularly when earth and Venus were in opposition. Bigg (1967) found an increase in sunspots and solar radio noise during times of mercury's closest approach. A doubling of this effect was noted at times when other planets aligned with mercury on the same side of the sun. These effects were generally assumed to be tidal in character, pulling of the sun's mass outward into a solar flare, producing radio noise. Wood (1972) has shown such influences on the sun when earth, Venus or Jupiter align. The tidal force of earth plus Venus is stronger than the force of Jupiter alone, as these planets are much closer to the sun. Gribbin and Plagemann (1974) use this effect as the basis for their popular book called "The Jupiter Effect", which relates possible solar-earth disturbances when all the planets will be in conjunction in 1982. On the basis of these affects, Cohen and Lintz (1974) and Wood (1972) predict 1982 as the coming sunspot maxima.

Nelson (1951) has often reported his method for prediction of earth's "radio weather" based on a heliocentric model of planetary position. His model, which reaches 93% accuracy, is based on angular relationships between two or more planets and the sun. The sun shows increased magnetic perturbation when any two planets align in either 0, 90, or 180 degree relationships. The most impressive disturbances are noted during alignment of an inner and outer planet. Sixty or 120 degree angles are associated with a calming effect upon solar radio weather.

Sleeper (1972) and Jose (1965) have shown the 180 year cycle of sunspots to be correlated with a resonance between the outer planets (Jupiter, Saturn, Uranus, Neptune), and the 11 year cycle to be correlated with a similar resonance between the inner planets (earth, mars, Venus, mercury).

Cohen and Lintz (1974) have similarly shown the 180 year solar cycle to be associated with a beat-frequency between the 9.8 and 11.0 year sunspot cycles. Sleeper (1972) has further demonstrated that the 9.8 year cycle is equal to 1/2 the synodic period of resonance of Jupiter-Saturn conjunctions, and, that increases in solar activity of up to 20% are associated with these times.

A Heavenly Body Influences the Weather on Earth

The relationship between solar phenomena and geophysical phenomena has been noted for some time. An increase in the number of sunspots seen on the solar surface generally coincides with:

1. Solar flare activity increase
2. Increased solar wind
3. Decreased terrestrial cosmic ray flux
4. Increased solar radiofrequency output
5. Increased terrestrial geomagnetic disturbance
6. Increased terrestrial auroral activity

The number of sunspots seen from earth varies from day to day but does have a distinct approximate 11 year periodicity. We have discussed in a previous section how the number of sunspots might be tied, in a highly predictable manner, to positions of the major planets relative to the solar surface. Additionally, there exist a 22 year cycle of magnetic polarity of sunspots, and a 5-6 year cycle, which is equal to 1/2 the regular 11 year cycle. All of these sunspot cycles have been related to earthly weather. The following tables summarize some of these effects:

Table II: EVENTS TIED TO THE 11 YEAR SOLAR CYCLE

1. Weather in North America - King (1973, 1974)
2. Circumpolar vorticity - Zatopek & Krivsky (1974)
3. Varve thickness & river stage measurements - Dewey (1970); Christy (1976).
4. Temperature changes in Leningrad, polar weather, planetary circulation patterns - Vitels (1957); Monin (1969)
5. Polar air pressure - Valdimirov, et al. (1970)
6. Storm frequency in Kara Sea - Mozalevskaya (1970)

Table III: EVENTS TIED TO THE 11 and 22 YEAR SOLAR CYCLES

1. Variations in climate - Rubenstein & Polzova (1966); Marshall (1972)
2. Air temperature - Polzova (1970)
3. River runoff - Galperin (1970); Davydova (1971)
4. Antarctic snow accumulation - Petrov, et al (1966)
5. Atlantic ocean temperature - Nikolaev (1970)
6. General Reviews - Rubashev (1964); Druzhinin & Khamyaova (1974)

Table IV: EVENTS TIED TO THE 22-YEAR SOLAR CYCLE

1. Worldwide pressure variations - Clayton (1920); Hanzlik (1951); Wagner (1971)
2. Worldwide alteration in temperature and precipitation - Willett (1968)
3. Rainfall cycling between central and northeastern U.S.A. - Marshall (1972); Roberts (1973)

Table V: EVENTS TIED TO THE 5-6-YEAR SOLAR CYCLE

1. Winds, precipitation, pressure, squalls, temperature and frequency of cold winters in England - Baxendell (1925)
2. Midlatitude temperatures - Hellman (1906)
3. Varve thickness, temperature and ice-free days in Neva, U.S.S.R. - Vkulis & Monin (1971)
4. European river runoff - Davydova (1971)
5. Eurasian river runoff - Druzhinin & Khamyaova (1974); Galperin (1970)
6. Atmospheric circulation - Haminov (1968); Bolotinskaya & Sleptzov (1964)
7. Wind patterns - Ohl (1972)
8. Air temperature - Baur (1953); Polozova (1970)
9. Precipitation - Drozdov (1970); Petrov, et al (1966)
10. Sea temperature - Nikolaev (1970)
11. Ocean ice cover - Rubashev (1964)

To summarize from the above cited works, and from other sources, we can identify the kinds of phenomena which associate with sunspot maxima (large number of sunspots) and sunspot minima (few sunspots):
(see Table VI next page)

Table VI: TERRESTRIAL SUNSPOT CYCLE REACTIONS

Variable	Sunspot Maxima	Sunspot Minima
Solar flares	increase	decrease
Terrestrial radio weather	disturbed	calm
Geomagnetic index	disturbed	calm
Cosmic ray flux	decrease	increase
Solar wind	increase	decrease
Barometric variability[6]	increase	decrease
Tropical temperatures[7]	decrease	increase
N. Atlantic storm tracks[8]	equatorward	poleward
W. coast U.S.A. storm tracks[9]	equatorward	poleward
Severe winters N.E. U.S.A.[10]	not likely	likely
Lightning[11]	increase	decrease
Solar "constant"[12]	increase	decrease
Tropical stratospheric wind reversals[13]	increase	decrease
Aleutian low[14]	fills in	deepens
Kazkhstan drought[15]	absent	present
Index of zonal circulation[16]	increase	decrease
Index of zonal and meridional circulation[17]	increase	decrease

---

6. Petersen (1945)
7. Kooper & Walker (1951)
8. ibid
9. Hansen & Tannehill (1951)
10. Baur (1953)
11. Stringfellow (1976)
12. Abott (1966); Kondratyev & Nikolsky (1970)
13. Schove (1973)
14. Maximov & Karklin (1970)
15. Baydal (1969)
16. Ped & Sidochenko (1959)
17. Katz (1960)

Onset of solar flares, terrestrial magnetic storms and auroral phenomena have also been related to the following meteorological phenomena:

Table VII: TERRESTRIAL SOLAR FLARE REACTIONS

1. Changes in storm tracks - Roberts (1973)
2. Increased mean east-west flow in midlatitude North America - Stolov & Shapiro (1975)
3. Changes in zonal atmospheric circulation - Loginov (1970)
4. Increased numbers of deep troughs, fewer minor troughs over N. Pacific - Roberts & Olsen (1973)
5. Increased vorticity over Gulf of Alaska - ibid
6. Perturbations in polar temperatures - Willett (1968)
7. Increased number of thunderstorm “whistlers” - Holzworth & Mozer (1977)

Similarly, the phenomena of solar magnetic sector passage has been associated with changes in winter low pressure trough area in the Northern Hemisphere (Wilcox 1973; Roberts 1973).

Lunar phase has likewise been linked to cycles of precipitation. Bradley, Woodbury & Brier (1962) and also Adderly & Bowen (1962) found significant cycles in rainfall occurring in phase with both full and new moon. Brier (1965) has further published evidence showing a 20% increase in precipitation just after syzygy.[18]

We can also note other interesting non-solar phenomena relative to an atmospheric principle of continuity:

1. Schonland (1930, 1933) found an increase in cosmic radiation below thunderstorms .

2. Fleisher and Falconer (1960) found “cosmic” radio noise similar to that of solar flares and stellar frequencies (18 mc/s) associated with thunderstorms.

3. An interesting relationship between cycles of barometric pressure and cloud streets has been reported by Landsberg and McGrath (1972).

4. A cycle in electrical effects termed as “spherics” has been isolated by Wedler (1973)

5. Nobel laureate Langmuir (1961), who coined the word “plasma” so widely in use today, noted a connectivity within the atmosphere not expected according to

---

18. When sun, moon and earth are all in a straight line, as during full or new moon periods.

classical meteorology. Cloudseeding experiments were performed on a seven day schedule in New Mexico. A similar seven day periodicity in precipitation followed in-phase, spreading from California to Western Europe, centered in the Ohio valley. Langmuir attributed this effect to an atmospheric "chain reaction", caused by excessive high altitude ice-crystal clouds formed during the seeding experiments.

6. In another context, Carson (1965) has reviewed anomalous meteorological data to conclude the existence of a continuum in the atmosphere.

7. A mechanism has also been suggested which would go against current ideas about the earth's balance of water. Interstellar hydrogen, existing in concentrations up to 1000 molecules/cm3 , could be entrained into the earth's atmosphere and combine with oxygen to yield water vapor. Such a mechanism has been postulated by McKay and Thomas (1978) with regard to galactic arm bands and long term climate cycles.

A Heavenly Body Influences the Earth's Inertia and Crustal Movement

Thus far we have related only the participation of the sun, planets and earth atmosphere in our cosmic symphony. Now we will discuss the moderation in tempo provided by participation of the geophysical structure of the Earth itself.

Suguwa (1973) has shown a strong correlation between solar activity, the atmosphere, and slight changes in the rotation of the earth. Anderson (1974) has further traced such variations in earth rotation to earthquake activity, as have Manshinha and Smylie (1970) and Dahlen (1971). Challinor (1971) has similarly traced fluctuations in the rotation of the earth (length of day) to both atmospheric and solar phenomena, as have Mintz and Munk (1951, 1954), Pines and Shaham (1973), and Vidorenkov (1968). Prins and Gillard (1971) have similarly found a relationship in the 2.25 year "chandler periodicity" earth wobble and equatorial wind patterns.

We are familiar with the ocean tides, however, both atmosphere and earth-crust rise and fall with the overhead passing of the moon and sun. Atmospheric tides are greatest according to solar position, while lunar position influences earth-crust and ocean tides the most. Highest tidal fluctuation for all three, ocean, air, and earth-crust, occurs during full and new moon periods (syzygy). Most recently, these interactions have been related to earthquakes.

Latham (1971, 1973a, 1973b) used seismometers set up by the Apollo astronauts to show that 80% of all moonquakes occur during the time of perigee or apogee. Tamrazyan (1967) has studied Nevada earthquakes and found an 800% increase within a few days after the moon's passage through perigee. Simkin (1973) has related a triggering effect of the sun and moon on earthquakes near Galapagos.

Myerson (1970), Pariyski (1970), Ryall (1968), Slichter (1972), and Shaw (1970) have all related the influence of earth tides to earthquakes. Myerson (1970) has further related this to earth wobble, while Whitten (1977) related earthquakes to polar motion.

Danjon (1959, 1960) and Schatzman (1966) relate that there is a link between solar flares and changes in the length of day. Gribbin and Plagemann (1973) have demonstrated the effect on earth spin from the solar storm of August 1972. The general problems involved have been discussed by Hirshberg (1972), McKinnon (1972), Rosenberg and Coleman (1974), Shatzman (1966), and Wilcox (1968).

IV. BIOLOGICAL, CHEMICAL, AND ELECTRICAL FACTORS

There is abundant evidence to show that living systems undergo cycles in activity and physiological parameters. Two schools of thought prevail in explaining the mechanism for such cycles. One school says the rhythms are caused by an internal mechanism which functions independent of the outside world, and the other says the rhythms are responses of the organism to subtle external environmental cues. Neither school of thought has been able to sway the all of the biologists as neither has been able to demonstrate the causal mechanism involved. The internal mechanism school believes the clock is somehow tied to the molecular rotations of a cell's DNA structure; the external mechanism school believes the clock is animal reaction to the waxing and waning of some environmental parameter, such as geomagnetism or gravity. The external or cosmic clock hypothesis will be related here. The cosmic clock researchers have, in my opinion, accumulated more evidence to support their case than the internal clock advocates.

This seems so, even though the majority of biologists currently adhere to the internal clock hypothesis. A reason for this may be the following: many of the cosmic timers are the heavenly bodies themselves. As we will discuss, some scientists back away from the cosmic biological clock hypothesis, claiming, unjustly, that it is somehow related to "astrology", as per a newspaper horoscope. This criticism is unfounded. Either an effect is associated with the position of a heavenly body or it is not, and the truth will never be determined so long as such labels stifle investigation and experiment. The material related here, the reader will discover, in no way compares with a newspaper horoscope.

Studies on the rhythmic nature of living systems extend back many centuries, and are as old as man's attempts to understand the cycles of planting and harvest, physiology and disease, and of the animal world in general. Petersen (1947) gives a good overview of the kinds of cycles in human functioning that exist and how some of them are timed to cycles of sunspots or to the weather.

Similarly, Luce (1970, 1973), Wolf (1962) and Palmer (1976) have covered the broad topic. Presman (1968) has gathered together evidence to show that the organism is sensitive to electromagnetic phenomena at natural ambient levels, as has Barnothy (1964) with regard to magnetic field phenomena. A science called biometeorology has grown over the years which is similar in scope to investigations of biological rhythms. While direct influences of weather upon an organism due to temperature, humidity or pressure are not our consideration here, an excellent overview of the field is given by Tromp (1963).

The biological clock is known to function independent of temperature, humidity or pressure, and other environmental parameters of a mechanical nature. When kept isolated in a laboratory under constant conditions, a wide variety of organisms are capable of "knowing" what the sun, moon, tides, etc. are doing outside of the controlled setting.

Knight (1972) has noted an effect of increased animal activity during times when the moon is directly "overhead or underfoot", corresponding to high tides near ocean areas, or to where a high tide would have occurred inland, if an ocean were nearby. Brown (1975) noted a similar periodicity in oyster and hamster activity related to lunar phase. He also demonstrated that oysters, when moved from Boston to Evanston, Illinois, would re-set their tidal activity cycle, of opening and closing, to the "tides" at the new location, even when kept under constant conditions. Brown further measured the metabolic rates of various organisms in hermetically sealed containers. Although kept under constant conditions of temperature, light, pressure, and humidity, they would show an "organismic knowledge" or reaction to local weather conditions and geophysical phenomena. (Brown 1975)

In these studies, geomagnetic influences and cosmic ray phenomena were suspected as the zeitgeber. These energies could penetrate the copper hermetic containers, while electromagnetic phenomena could not. Brown's studies of metabolic rates show a high correlation with primary cosmic ray activity, and an interesting ability for certain species to predict the barometric pressure two days into the future,[19] even when kept in hermetically sealed containers. The implication here is that the organism responds to some factor in the environment which might be a major determinant force for steering frontal movements; in other words, the sensitive organism responds now to a factor in the environment which the atmosphere responds to in a much slower fashion.

---

19. The sunflower, Silphium laciniatium, will indicate the points of the magnetic compass; Indian licorice, Arbrus precatorius, is sensitive to electrical fields and is nicknamed the weather plant, stemming from its historical use to predict tornadoes, hurricanes, earthquakes, and volcanic eruptions (Tompkins 1973). We are also reminded here of wild animal precognitions of earthquakes, floods, etc. often well in advance of their occurrence.

A large number of organisms have shown responsiveness to magnetic fields at levels no stronger than the natural earth field. Brown (1975) has summarized previous research which shown responses of honeybees, flies, robins, homing pigeons, gerbils, hampsters and other organisms to magnetism. Brown also demonstrated organismic response to gamma radiation at 2 - 25 times the natural background level. Similarly, a two year study showed a cycle of potato metabolism spatially oriented to certain areas of the galaxy. Throughout the sidereal day, maximum oxygen consumption occurred with the constellation Gemini overhead, while minimum oxygen consumption occurred with Sagittarius overhead. It is not that these constellations bear any mystical significance to oxygen consumption, but they signify a spatial orientation between the earth's axial tilt, the plane of the solar system, and the plane of our galaxy, which periodically realign geocentrically, each sidereal day. Our three dimensional orientation to centers of energy activity in space becomes the prime factor for understanding any astrophysically-linked cycle in organismic or meteorological parameters. The resolution of such considerations are classic geographic problems, involving pattern recognition and spatial orientation in three dimensions, rather than the usual two.

In another interesting context, Brown (1975) identified cycles of water uptake in bean seedlings which possessed a lunar periodicity. Maximum water uptake occurs at full and new moon periods. The old farmer's lore is thereby vindicated: not only does the farmer have more light to work by during the busy planting period, but full moon planting assures a more active seedling, plus a greater chance of precipitation!

Regarding human biological responses, rapid fluctuations in weather, as in alteration of warm and cold fronts, appears to either depress or stimulate certain ill persons. In such cases the predisposition of the individual seems to be the determining factor, with the weather acting as a triggering mechanism to act upon latent infections, scar tissues, or upon the sympathetic nervous system. "Pains due to the weather" are often reported long before actual measured arrival of weather phenomena. We can list some of the more significant findings regarding such living system interaction with the weather:

1. Link and DeRudder (1947) have shown a correlation between diseases and weather fronts, prefrontal subsidence and orographic foehn. The foehn winds of the alps and the Chinook winds of the eastern Rocky Mountains are caused by a rush of subsiding air, warming and raising the pressure in a particular area. These winds have been associated with headache, nausea, sleeplessness, lethargy, irritability, a general feeling of debilitation, and an overall increase in pulse rate with a corresponding drop in blood pressure.

2. Petersen (1947) determined that fronts have different effects upon human physiology in general: cold fronts being related to anabolism (tendency toward complex molecular structuring), and warm fronts being related to catabolism (tendency toward simple molecular structuring). Curry (1947) classified patients according to individual sensitivity to warm or cold frontal activity.

3. Petersen (1947) also noted fluctuations in blood pH and white blood cell count to be associated with passage of frontal systems: cold fronts increase white cell count and decrease blood pH, while warm fronts have the exact opposite effect.

4. Bortels (1969) found an inhibited microorganism activity associated with warm fronts, and increase in such activity with cold frontal passage. Such responses were noted even when the microorganisms were kept under constant environmental conditions. Bortels attributes these effects and similar ones as due to a specific "weather radiation" which possesses a dual character, one type from cyclonic systems and the other type from anticyclonic systems. Bortels further noted similar effects upon chemical, non-living systems, as we shall mention shortly.

5. A relationship between weather, childbirth and sex of offspring has also been noted (Petersen 1947). Increases in temperature coincide with a decrease in numbers of male infants born. In a curiously similar fashion, more males die during passage of a cold wave; more females die during a heat wave. Males seem to be more effected by cold and females by heat, with respect to both time of conception and mortality.

6. We also note that during passage of a cold front there is a greater admission rate of psychotic patients at hospitals (Petersen 1947). More schizophrenic episodes occur during spring and cold weather, while manic depressive are affected most during autumn and summer heat. The connectivity of psychiatric problems with the environment dates back to the work of Spiegelus (1627), who noted a swelling of the brain during the full moon. Hunter (1835) further elaborated on the connection between "mad" people and the full moon. More recently, Beamer-Maxwell and Hendricks (1951) noted an increase in psychiatric admissions during new and full moon, as did Pokorny (1962). Pokorny (1964), Taylor and Diespecker (1972) have noted increases in suicides during new and full moon; Olivenson, Levinson and Lester (1942) have also noted an increase in suicides around full moon. Lieber and Sherin (1972) have demonstrated that increased murders take place during full moon periods.

7. Arrhenius (1898), and later Koopen, Richter and Elkholm (1947) noted peaks of scarlet fever, epileptic seizures, mortality and sex ratio of newborn associating with full moon periods. Petersen (1947) and Berg (1947) made similar connections. Petersen also showed that overall mortality roughly followed the index of barometric variability (instability). Both mortality and instability show

a coordinated monthly cycle and a three year cycle. Dull (1937, 1939) has related sunspot activity to neuroarthritis (weather pains) and to suicides.

8. Tchijevsky (1930, 1964, 1971) and Wheeler (1951, 1960), in separate studies, have found cyclic relationships in wars and civil disturbances, migrations and epidemics, all tied into cycles of solar and lunar periodicity. Tchijevsky noted a 72% increase in such activity at times of major solar flares. Disease symptoms and latent mass anger seem to be pushed to the surface during solar eruptions.

9. Burr (1972) measured electrical potentials of several oak trees for more than a decade. Fluctuations in potential gradients of two trees commonly recorded activity of approaching storms well in advance of the actual outbreak of the storm. Storm intensity could similarly be predicted from electrical potential measurements. Burr's basic discovery was that of a ubiquitous electrical field, termed the "electrodynamic field". Fluctuations in this field, measured as electrical potential differences on the order of volts, were found in all living organisms and in the earth crust, atmosphere, and other inanimate materials. Interestingly, while the base levels for such electrical potential measurements are different for each object or organism being measured, all objects and organisms within a given local area (trees, animals, rocks, atmosphere) will show identical simultaneous positive upswing or negative downswing in electrical potentials. These swings of potential also show lunar and solar periodicities, indicating a common environmental agent ruling electrical parameters in a given environment. Burr felt that this field phenomena ruled molecular interactions, in both the living and non-living realms. Changes in the electrical patterns would always precede and predict changes in either tissues or weather.

10. Piccardi (1962) has shown that seemingly isolated quantitative chemical interactions can be influenced by the dynamic character of its surroundings. Piccardi showed that precipitation of various chemical substances, notably bismuth chloride, would vary according to the daily sunspot number. His experiments, as well as those of Brown, coordinate with the sidereal earth-orientation in space relative to the plane of the galaxy. Most interestingly, Piccardi found that by shielding the experiment with metal plate or with a Faraday cage, a new parameter was introduced. Some experiments were noticeable altered by such an enclosure, while cyclic components were either amplified or eliminated.[20]

11. Bortels (1956, 1965) similarly showed cycles of weather activity to correlate with "isolated" chemical interactions, leading him to speculate on the existence of a radiation emanating from weather systems, a different type for cyclonic and

20. These factors strongly relate to the previously mentioned ether-drift experiments, and to Reich's orgone energy experiments.

anticyclonic weather. Bortels further showed such a chemical activity to match effects upon isolated microbial cultures. Both microbes and supercooled water samples, even when kept in "controlled" laboratory conditions, would show cycles of activity, or amount of freezing, dependent upon whether barometric pressure was rising or falling. Placing the experimental sample inside metal screens or foils would change the character of the effect. Solar phenomena could moderate the effect as could lunar position. Bergter and Noack (1956) have confirmed these effects. Bortels (1965) has also shown an effect upon microbes and freezing of supercooled water which can be modified by location of experiment. Attic locations showed distinct differences as compared to basement locations. In a similar vein, Rau (1951) has shown that the freezing of supercooled water can be triggered by strong electric fields.

V. SUMMARY

The following three tables roughly summarize the phenomena related in the previous sections:

Table VII: WEATHER INDUCED CHANGES

Variable	Cold Fronts	Warm Fronts
Basal metabolic rate	high	low
Blood pH	low	high
White blood count	high	low
Microorganism activity	increases	decreases
Psychiatric episodes	schizophrenic	depressive
Sex-ratio of offspring	more males	more females
Sex-ratio of dead	more males	more females
tendency toward	anabolism (complexity)	catabolism (simplicity)

Table VIII: LUNAR INDUCED CYCLES

Variable	Syzygy	Non-Syzygy
Precipitation	increase	decrease
Bean water uptake	increase	decrease
Animal activity	increase	decrease
Tides (ocean, earth-crust,	increase	decrease
atmosphere)	increase	decrease
Earth electrical potential	increase	decrease
Psychiatric episodes	increase	decrease
Crime rate	increase	decrease
Suicides	increase	decrease
Epidemics	increase	decrease
Epileptic seizures	increase	decrease

Table IX: SOLAR INDUCED CHANGES (includes material in Table VI)

Variable	Sunspot Maxima	Sunspot Minima
Solar flares	increase	decrease
Terrestrial radio weather	increase	decrease
Geomagnetic disturbance	increase	decrease
Cosmic ray flux	decrease	increase
Solar wind	increase	decrease
Barometric variability	increase	decrease
Tropical temperatures	decrease	increase
Storm tracks, N. Hem.	equatorward	poleward
Severe winters, N.E. U.S.A.	decrease	increase
Lightning	increase	decrease
Solar "constant"	increase	decrease
Tropical stratospheric wind reversals	increase	decrease
Pressure in Aleutian low	increase	decrease
Kazkhstan Precipitation	increase	decrease
Zonal & meridional circulation index	increase	decrease
Earth electrical potential gradient	increase	decrease
Freezing of supercooled water	enhanced	suppressed
Chemical precipitation rates	increase	decrease
General mortality	increase	decrease
Epidemics	increase	decrease
Microbe virulence	increase	decrease
Psychotic episodes	increase	decrease
Epileptic attacks	increase	decrease
Maleness of dead (wars?)	increase	decrease
Precipitation	increase	decrease
River stages, varves	increase	decrease

We put these tables together without comment. The reader should be able to note a number of distinct, interwoven patterns, with a general unifying structure. A full theoretical integration of these tables with more conventional natural-scientific observations will be presented at a later time under a different context.

We have come full circle, showing a link of interrelated phenomena connecting the worlds of biology, chemistry, physics, astrophysics, and meteorology. These disciplines are but interconnected facets of one single universe, separated only in the minds of humans and maintained, as such, by our incomplete knowledge and faulty perceptions. The earth functions as a whole, and is "plugged" into the cosmos as well. The major mystery, as this point, is the essence of the energetic, cosmic glue and driving force, which organizes, moves, and binds it all together.

Appendix Postscript, Added 2010

The above Appendix material argues for the existence of an *interconnecting principle in the natural world*, the "Principle of Atmospheric Continuity" which in fact permeates all of the universe, in the nature of a cosmic ether in space, or of Reich's cosmic orgone energy. I took great risks to include such material into the Thesis, given it's heretical nature; but without it, the whole argument that the cloudbuster could actually have a quick influence over tens to hundreds of miles of distance would have rang hollow. If such an energy principle existed, so the rational critic could argue, then it must have some expressions in nature which other scientists besides Wilhelm Reich would have detected, even if they described it somewhat differently and gave it completely different names. And indeed, that was the case. The Appendix was therefore prepared and presented from materials I had been gathering from my various readings over the prior years, going back to my undergraduate days.

However, there is more to it than I could easily elaborate upon in the Thesis. The Tables given in the Summary section (Tables VII, VIII and IX), may in fact be considered as evidence for a *Basic Antithesis of Bio-Energetic Functions in Nature*. This is a concern which harkens back to Reich's early research as a medical psychoanalyst. In the 1930s, he identified a basic expansive-contractive life-energetic pulsation within the entire human biosystem, which affected the totality of one's being. Energetic currents moved from core to periphery and periphery back to core, which he found expressed in general states of pleasure (expansion) and anxiety (contraction). These energetic movements fundamentally affected the whole complex of emotions, psyche, somatic and body-posture reactions. From these clinical observations, Reich argued for a *Basic Antithesis of Vegetative Life Functions* (see his book *Bioelectric Investigation of Sexuality and Anxiety*) which were foundational for his later discovery of the life-energy, or orgone energy. Today we know the free life-energy in the atmosphere and in high-vacuum also has expansive-contractive expressions, suggesting the same plasmatic functions as seen in the living organism are found extending throughout the non-living world.

What I argue here is, the same pulsatory energy principle extends into non-living nature, on the planetary scale and throughout the entire Solar System, possibly functioning within galaxies. Our Sun occupies the central-most core of the Solar System, with its own field of energy composed, in classical terms, of magnetic, electrical and ionic components. The plasma physicists have termed this the *cosmic plasma* of space, which has very life-like motions and behaviors, as one sees in high-altitude aurora phenomenon. It exists everywhere, much like the older concept of *cosmic ether,* except in that *it pulsates, expands and contracts.* From Reich's concepts, we may view this energy-rich cosmic plasma-ether as extending outwards from the Sun, composing an energy field surrounding the Sun and expanding well out into planetary space whereupon it merges into the background interplanetary medium. The cosmic energy field of the Solar System pulsates over very long cycles, of "breathing" out with expansion during years of high sunspots and solar flares, with a strong solar wind. In the reverse phase of the pulsation cycle, there is a "breathing" in, or reduced outflow of the Sun's activity, in a contractive phase. Various solar and planetary features respond to these long-duration solar pulsations, which can affect the weather, climate, chemistry and biology on our small planet, and the other planets as well. The clinical health worker, biochemist, psychologist and atmospheric scientist who follows the lunar and solar phases will know about these periodicities. The expansive-contractive pulsations of our Sun are similar to a jellyfish pulsation, bioenergetically speaking, with cycles lasting over tens to hundreds of years. The pulsations affect the full Solar System in systematic and predictable ways.

Frequent sunspots and flares create a kind of overexcited and expansive high-energy condition within the solar system. Auroras blossom into brilliance, with shifts and adjustments in the Earth's geological dynamics and weather systems, along with human biosocial reactions, and other reactions in animal life and even raw inorganic chemistry. When the solar-terrestrial configurations reverse, marked predominantly by a lowering of sunspot abundance, an antithetical set of reactions develop, across the board.

A similar set of expansive/contractive, high-low excitation reactions also occur during the normal progress of the Lunar cycle, which affects ocean tides, and has additional expressions in the atmosphere and Earth's crust, as well as in biological systems. Full alignment of the Sun-Earth-Moon system (syzygy) creates an added though more temporary and short-cycle excitation in the Earth's energy field, with a set of expansive influences therefrom. Additional pulsation-excitation influences develop from the progress of weather systems over the Earth's surface, high versus low pressure, or dry versus wet, and these go well beyond the more mechanical influences of only Lunar brightness, or temperature and rain, or humidity and pressure patterns associated with changing weather. The scientists studying natural cycles have already proven an abundance of correlated patterns, even if the deeper-lying orgonotic, life-energetic mechanism, of pulsation and excitation, has eluded them.

While much of the exacting details has yet to be worked out, in this Appendix I have made a preliminary attempt at identifying the larger patterns. Obviously, the effort undertaken in 1979 could be adjusted and expanded upon with new knowledge acquired by 2010. For now, it is merely noted that such large cosmical-bio-meteorological patterns exist, and require some kind of unifying energy principle in nature which parallels what is necessary for Reich's cloudbuster apparatus to function. They are expressions of the larger expansive-contractive energy system of our Sun and Solar System. And, they pulsate in the general directions as I have indicated in the given Tables. A more complete discussion on this interesting subject will be forthcoming.

LITERATURE CITED (Appendix)

Abbot, C.G. (1966): "An Account of the Astrophysical Observatory of the Smithsonian Institution, 1904-1953", *Smith. Misc. Coll.* '4656, 148(7).

Adcierly, E.E. & Bowen, E.G. (1962): "Lunar Component In Precipitation Data", *Science*, 137:749-750.

Anderson, D. (1974): "Earthquakes and the Rotation of the Earth", *Science*, 186:49

Anderson, LJ.L. & Spangler, G.W. (1973): "Serial Statistics: Is Radioactive Decay Randolm?", *Journal of Physical Chemistry*, 77:3114-21.

Arp, H., Bahcall, J .N. & Field, G.B. (1973): *The Redshift Controversy,* W.A. Benjamin, Reading, MS.

Arrhenius, S. (1898): "Die Einwirkung Einflusse auf Physiologische Verhaltnisse", *Skandinavishe Archiv fur Physiologie*, 8:367-416.

Asimov, 1. (1966): *The Neutrino*, Avon Books, New York.

Baker, C.F. (1977): "The Pendulum Experiment", *Journal of Orgonomy*, 11:176-187, New York

Barnothy, M.F. (1964): *Biological Effects of Magnetic Fields*, Plenum Press, New York.

Baur (1949): Contained in *Compendium of Meteorology*, American Meteorological Society, Boston (1953).

Baxendell (1925): "Meteorological Periodicities of the Order of a Few Years", *Quart. J. Roy. Meteor. Soc.*, 51:371-392.

Baydal (1969): "About the Influence of Solar Activity on the Atmospheric Precipitations", *Trudy Kazakhsk. Nauchno-issledov*, Gidromet. Inst. 38:22-24.

Beamer-Maxwell, E. & Hendrick, H.F.M. (1953): Quoted in Ravitz (1953): "Electrodynamic Field Theory in Psychiatry", *Southern Medical Journal,* 46:650-660.

Berg (1947): contained in Petersen (1947).

Bergter, F. & Noack, D. (1956): "Uber eine auf biologische und Physiko-chimische Systeme Wirksame Komponente der kosmischen Strahlung (KS) ", *Naturwissenshaften*, 43:208.

Berkson, J. (1966): "Examination of Randomness of Alpha Particle Emissions", in *Research Papers in Statistics*, F.N. David, ED., Wiley, NY.

Bigg, E.K. (1967): "Influence of the Planet Mercury on Sunspots, *Astronomical Journal,* 72:463.

Bolotinskaya & Sleptzov (1964): "About the Influence of Solar Activity on... Atmospheric forms", *Problemy Arktiki i Antarkt.*, 18:48-56.

Bortels, V.H. (1951): "Beziehungen zwischen Witterungsablauf, physikalisch-chemischen Reaktionen, biologischem Geschehen und Sonnenaktivitat", *Naturwissenschaften*, 38:165-176.

———— (1956): "Die hypothetische Wetterstrahlung als vermutliches Agens kosmo-meteorobiologischer Reaktionen", *Wissenschaftliche Seitschrift der Humboldt-Universitat zu Berlin*, VI:115-124.

———— (1965): "Das Gefrieren kleiner Wassermengen als solar-meteoro-biologische Modellreaktion", *Naturwissenschaften*, 52:118.

———— (1969): "Meteorologische Einflusse auf Atmungsund Garungsintensitat von Saccharomyces Hefe", *Naturwissenschaften*, 56:574.

Brace, D.B. (1904): "On Double Refraction in Matter Moving through the Aether", *Phil. Mag.* 6th series, 7:317-329.

Bradley, D.A., Woodbury, M.A. & Brier, G.W. (1962): "Lunar Synodical Period and Widespread Precipitation", *Science*, 137:748-749.

Brier, G.W. (1965): "Diurnal and Semidiurnal Atmospheric Tides in Relation to Precipitation Variations", *Monthly Weather Review*, 93:93.

Brown, F.A. (1975): "Evidence for External Timing of Clocks", contained in: *An Introduction to Biological Rhythms*, J. Palmer, Ed., Academic Press, New York.

Brown, T.T. (1960): "Electrokinetic Apparatus", U.S. Patents # 2,949,550; # 1,974,483; #3,022,430; 4/: 3,187,206.

Burr, H.S. (1972): *The Fields of Life*, Ballantine Books, New York.

Carson, R.B. (1965): *On the Nature of the Elongated Convective Belt*, private publication, Miami, FL.
Challinor, R.A. (1971): "Variations in the Rate of Rotation of the Earth", *Science*, 172:1022.
Christy, D. (1976): "Sunspot Polarity Vs River Flow", *American Society of Agricultural Engineers*, University of Nebraska.
Clayton, H.H. (1920): *World Weather,* Macmillan Co.
Cohen, T.J. & Lintz, P.R. (1974): "Long Term Periodicities In the Sunspot Cycle", *Nature*, 250:398.
Conklin, E.I<. (1969): "Velocity of the Earth with Respect to Cosmic Radiation", *Nature*, 222:971.
Curry (1947): contained in Petersen (1947).
Curtiss (1932): contained in Anderson & Spangler (1973).
Dahlen, F.A. (1971): "The Excitation of the Chandler Wobble by Earthquake", *Geophysical Journal*, 25:157.
Danjon, A. (1959): "Solar Flares and Changes in the Length of the Day", *C. R. Acad. Sci.* series B, 249:2254.
—————— (1960): ibid, 250:1399.
Davydova (1971): "About Spectral Methods of Hydrological Data Investigations", *Meteorologia i Gidrologia*, 3:72-79.
deBroglie, L. (1960): *Non-Linear Quantum Mechanics*, (translation), Elsevier, New York.
deGrazia, A. (1966): *The Velikovsky Affair*, University Books.
DePalma, B.E. (1976): "The Effect of Gravity on Rotating Objects", "Gravity and the Spinning Ball Experiment", and "Simple Experimental Test for the Inertial Field of a Rotating Real Mechanical object", contained in: *Is God Supernatural?*, Dione, R.L., Bantam Books, New York.
Dewey, E.R., Ed. (1970): *Cycles: Selected Writings*, Foundation for the Study of Cycles, Pittsburgh, PA.
—————— (1971): *Cycles: Mysterious Forces that Trigger Events*, Hawthorn books, New York.
—————— (1977): "Letter to Members", *Cycles*, Vol. 28, #3, p. 51.
Dirac, P.M. (1951): "Is There An Ether?", *Nature*, 162:906.
Ditchburn, R.W. & Heavens, O.S. (1952): "Relativistic Theory of a Rigid Body", *Nature*, 170:705 c f. pp 243-244.
Dudley, H.C. (1959): *New Principles in Quantum Mechanics*, Exposition University Press, NY.
—————— (1976): *Morality of Nuclear Planning?*, Kronos Press, Glassboro, NJ.
Dull (1937): contained in Petersen (1947) (1939): ibid.
Drozdov (1970): "About the Stability of Atmospheric Precipitation Cycles", *Trudy Glavnoi Geofis. Observ.*, 258:64-80.
Druzhnin, LP. & Khamyaova, N.V. (1974): "Solar Activity and Sudden Changes in the Natural Processes on Earth — A Statistical Analysis", NASA Trans. TTF-652 (NTIS document TT 70-57264).
Einstein, A. (1905): "Uber die von der molekularkinetischen Theorie der Warme geforderte Bewegung von in ruhenden Flussigkeiten suspendierten Teilchen", *Annalen der Physik,* 17:549-560.
—————— (1916): "Relativity: the Special and General Theory", (trans.) Holt, New York, 1920.
Emery, G.T. (1972): "Perturbation of Nuclear Decay Rates", contained in *Annual Review of Nuclear Science*, Annual Reviews, Inc., Palo Alto, CA.
Essen, L. (1955): "A New Aether-Drift Experiment", *Nature*, 175:793-794.
Fleischer, R. & Falconer, R.E. (1960): "Cosmic Noise and the Atmosphere", *Obs. of Rensselaer Polytech. Inst.*, Troy, New York, p. 191-207.
FOx, K. & Jennings, D. (1977): as reported in *Science News*, Dec. 10, 1977, V. 112.
Galperin (1970): "About the Cycles in Climatic Patterns in Kazakhstan", *Sbornik rabot Alma-atinskoi Gidromet. Observ.*, 5:164-171.

Gribbin, J. & Plagemann, S. (1973): "Discontinuous Change in Earth's Spin Following Great Solar Storm of August, 1972", *Nature*, 243:26.
——————————— (1974): *The Jupiter Effect*, Randolm House, NY.
Haminov (1968): "Solar Activity... Atmospheric Processes", *Izvestiya Uses Geodgraf. Obsheh.*, 2:103-108.
Hansen & Tannehill (1951): contained in *Compendium of Meteorology,* American Meteorological Society Boston, MS.
Hanzlick (1951): ibid Hellmann (1906): "Die Niederschlage in den Norddentschen Strongebieten", Berlin, I.
Hirshberg, J. (1972): "Upper Limit of the Torque of the Solar Wind on the Earth", *J. of Geophysical Research*, 77:4855.
Hunter, J. (1835): Works, Longman Rees & Co., London Holzworth & Mozer (1977): contained in *Science News*, Vol. Ill, 18 June 1977, p.389.
Illingworth, K.K. (1927): "A Repetition of the MichelsonMorley Experiment Using Kennedy's Refinement", *Physical Review*, 30:692-696.
Jose, P.D. (1965): "Sun's Motion and Sunspots", *Astron. J.*, 70:193.
Juergens, R.E. (1976): "Plasma in Interplanetary Space: Reconciling Celestial Mechanics and Velikovskian Catastrophism", in *Velikovsky Reconsidered,* Doubleday & Co., NY.
Katz (1960) : "The Seasonal variations of General Atmospheric Circulation and Long Range Weather Forecasting", *Gidrometeorizdat*, Leningrad, p.270.
Kennedy (1935): "Simplified Theory of the MichelsonMorley Experiment", *Physical Review*, 47:965-968.
King, J.W. (1973): "Solar Radiation Changes and the Weather", *Nature* 245:443.
King, J.W. (1974): ibid., *Nature*, 274:131.
Kondratyev & Nikolsky (1970): "Solar Radiation and Solar Activity", *Quart. J. Roy. Met. Soc.*, 96:509-522.
Kooper & Walker (1951): contained in *Compendium of Meteorology*, Am. Met. Soc., Boston, MS.
Koopen, Richter and Elkholm (1947): contained in Petersen (1947).
Koyzrev, N.A. (1968): *Possibility of Experimental Study of the Properties of Time*, JPRS Translation #45238.
Kubrin, D. (1972): *How Sir Isaac Newton Helped Restore Law'n Order to the West,* privately published.
Kurzner, W. (1924): "The Exception", *z. Phys.*, 21:281; also (1927): ibid., 44:655.
Landsburg, H.E. & McGrath, P.A. (1972): "Periodicities in Daily Records of Global Radiation at College Park, Md., USA", *J. Interdicipl. Cycle Res.*, 3:91-98.
Langmuir, I. (1961): *The Collected Works of Irving Langmuir, Vol. 10: Atmospheric Phenomena; Vol. 11: Cloud Nucleation*, Pergamon Press, New York.
Latham, G. (1971): "Moonquakes", *Science*, 174:687.
————— (1973a): *Science*, 181:49.
————— (1973b): "The Moon", *Science*, 7:396.
Lieber, A.L. & Sherin, C.R. (1972): "Homicides and the Lunar Cycle", *Am. J. Psychiatry*, 129:69-74.
Link & deRudder (1947): contained in Petersen (1947) 200.
Loginov (1970): "About the Manifestations of Different Solar Disturbances on the Atmospheric Circulation", *Trudy Glavnoi Geofizich. Observ.*, 258:18-23.
Lorentz, H.A. (1889): "De L'influence du Mouvement de la Terre sur les Phenomenes Lumineux", *Arch. neerlandaises des sciences exates et naturelles,* 21:103-176.
Luce, G.G. (1970): *Biological Rhythms in Psychiatry and Medicine*, N.I.M.H., D.H.E.W., pub #2088, Govt. Print. off., Wash. DC.
————— (1973): *Body Time*, Bantam Books, NY.
Mansinha, L. & Smylie, D.E. (1970): "Seismic Excitation of the Chandler Wobble" in *Earthquake Displacement Fields and the Rotation of the Earth*, Reidel, Holland, p.112.
Marshall, J., (1972): *Precipitation Patterns of the United States and Sunspots*, Dissertation, U. of Kansas, Geography-Meteorology Dept., Lawrence, KS.

Maximov & Karkland (1970): "Seasonal and Long-term Variations of Depth and Position of the Aleutian Low, 1899 - 1954", *Izvestiya Vses. Geograf. Obsheh.*, 102, 5:422-431.
McKay & Thomas (1978): contained in *Science News*, undated.
McKinnon, J.A. (1972): *August 1972 Solar Activity and Related Geophysical Effects*, N.O.A.A. Tech. Mem. ERLMSEL-22.
Merritt, R.F. (1974): "Theories About the Properties of Rotation", *Ufo Quarterly Review*, April-June, p. 8-23, NICAP.
Michelson, A.A. (1882): "The Relative Motion of the Earth and the Luminiferous Ether", *Am. Journal of Science*, 23:395.
——————— (1927): contained in Swenson (1972).
Michelson, A.A. & Morley, E.W. (1887): "On the Relative Motion of the Earth and the Luminiferous Aether", *Am. J. of Sci.*, series 3, xxxiv:203.
Miller, D.C. (1921): contained in Miller (1933).
——————— (1933): "The Ether-Drift Experiment and the Determination of the Absolute Motion of the Earth", *Reviews of Modern Physics*, 5:203-42.
Mintz, Y. & Munk, W.H. (1951): "The Effect of Winds and Tides on the Length of the Day", *Tellus,* 3:117.
Mintz, Y. & Munk, W.H. (1954): ibid, *Geophys. Supp. Mon. Note Roy. Ast. Soc.*, 6:566.
Monin (1969): "Weather Forecasting as a Physical Problem", *Izdatelsto 'Nauka'*, Moscow, p.184.
Morley, E.W. & Miller, D.C. (1904): Letter to Lord Kelvin, in *Philosophical Magazine*, series 6, 8:753-754.
————————————— (1905): ibid., 9:669-680.
Mozalevskaya (1970): "On the Relationship between Storms in the S.W. Kara Sea and Solar Activity", *Problemy Arktiki i Antarkt*, 34:85-89.
Myerson, R.J. (1970): "Evidence for Association of Earthquakes With Chandler Wobble, Using Long Term Polar Data of the ILS-IPMS", in *Earthquakes, Displacement fields and the Rotation of the Earth*, Manshinha, Smylie & Beck, Ed's., Reidel, Holland.
Nelson, J.H. (1951): "Circuit Reliability, Frequency Utilization and Forecasting in the High Frequency Communications Band", in *Effects of Disturbances of Solar Origin on Communications*, Glassmann, Ed., Macmillan, NY.
——————— (1951): ibid, *RCA Review*, March, p. 26.
——————— (1954): ibid, *Trans. IRE* CS-2, No. 1:19.
Nikolaev (1970): "A Search for Hidden Periodicities in Many Dimensional Values", *Trudy Arktich. i Antarktich. Inst.*, 296:141-150.
Ohl (1972): "Solar Activity Cycles and their Geophysical Manifestations", *J. Interdisciplinary Cycle Research*, 3:395-408.
Oliven, Levinson & Lester (1942): "Moonlight and Nervous Disorders", *Am. J. of Psychiatry*, 99:579-584.
Palmer, J. (1975): *An Introduction to Biological Rhythms*, Academic Press, NY.
Pariyski, N.N., et ale (1970): "On the Problem and the Role of Trigger Mechanisms in Earthquakes", *Akad. Nauk USSR-Inst. Geophysic*: 62.
Ped & Sidochenko (1959): "On the Cause of the Intensity of Variations of Zonal Atmospheric Circulation", *Meteor. i Gidrologiya*, 9:20-24.
Petersen, W.F. (1947): *Man, Weather, Sun*, Chas. Thomas, Springfield, IL.
Petrov, et ale (1966): "Periodicity of Snow Accumulation in the Antarctic", *Inform. Bull. Sovetskoi Antarkticheskoi Exped.*, 57:97-106.
Piccard, A. & Stahel, E. (1926): "L'Experience de Michelson realisee en ballon libre", *Comptes Rendus,* 183:421.
Piccardi, G. (1962): *Chemical Basis of Medical Climatology*, Chas. Thomas, Springfield, IL.
Pines, D. & Sham, J. (1973): "Seismic Activity, Polar Tides and the Chandler Wobble", *Nature*, 245:77.
Pokorny, A.D. (1962): "Moon Phases and Mental Hospital Admissions", *J. of Psych. Nursing,* 6:325-327.
——————— (1964): "Moon Phases, Suicide and Homicide", *Am. J. of Psychiatry*, 121:66-67.

Polozova (1970): "Analysis of the Cyclical Oscillations of the Mean Monthly Air Temperatures in the Northern Hemisphere", *Trudy Glavnoi Geofizich. observ.,* 269:36-73.

Presman, A.S. (1970): *Electromagnetic Fields and Life*, Plenum Press, NY.

Prins & Gillard (1971): "Study of the Rotation of the Earth", *Interdisciplinary J. of Cycle Research*, 2,3:325-329.

Rau, W. (1951): "Eiskeimbildung durch dielektrische Polarisation", *Naturforsch*, 6a, 649-657.

Ravitz, L.J. (1962): "History, Measurment, and Applicability of Periodic Changes in the Electromagnetic Field in Health and Disease", *Annals, N.Y. Academy of Sciences*, 98:1144-1201.

Rayleigh, Lord (1904): "Does Motion Through the Aether Cause Double Refraction?", *Phil. Mag.*, series 6, 4:678-673.

Reich, W. (1948): *Discovery of the Orgone, Vol. 2: The Cancer Biopathy*, Orgone Institute Press, NY.

——— (1951): *The Oranur Experiment, First Report: 1947-1951, Orgone Institute Press*, NY.

——— (1957): *Contact With Space: Oranur Second Report 1951-1956*, Core Pilot Press, NY.

Righi, A. (1919): "L'Experience de Michelson st son interpretation", *Comptes Rendus*, 168:834 cf. (1920): ibid, 170:1550-1554, and (1920): ibid, 171:22-23.

Roberts, W.O. (1973): "Does Variable Solar Activity Affect Weather or Climate?", *Trans. AGU,* 54, 4:444.

Roberts, W.O. & Olsen, R.H. (1973): "Geomagnetic Storms and Wintertime 300-mb Trough Development in the North Pacific-North Atlantic Area", *J. Atmos. Sci.,* 30:135; cf. (1973): ibid, *Science*, 180:185, and (1973): ibid, *Rev. Geophysics & Space Physics*, 11:731.

Rosenblum, F. (1970): "The Red Shift", *J. Orgonomy*, 4:183.

Rosenberg, R.L. & Colman, P.J. (1974): "27-Day Cycle in The Rainfall at Los Angeles", *Science*, 250:481.

Rubashev (1964): "Problems of Solar Activity", Ch.1 Sect.3, NASA Tech. Trans., F-244, Wash. DC.

Rubenstein & Paolzova (1966): "Contemporary Climatic Changes", *Gidrometeoizdat*, Leningrad, p.268.

Ryall, A., et al. (1968): "Triggering of Microearthquakes by Earth Tides", *Bull. Seis. Soc. Am.*, 58:215.

Sagnac, M.G. (1913): "L'Ether lumineux Demonstre par l'effet du vent relatif d'ether dan interferometre en rotation uniforme", *Comptes Rendus,* 157:710; cf. Chappell, J.E. (1965): "Georges Sagnac and the Discovery of the Ether", *Arch. internat. d'histoire des Sc.*, 18:175-190.

Saxl, E.J. (1964): "An Electrically Charged Torque Pendulum", *Nature*, July 11, 203:136-138.

Schatzman, E. (1966): "Interplanetary Torques", in *The Earth-Moon System*, Cameron & Marsden, Eds., Plenum Press, NY.

Schonland, B.F.J. (1930): "Thunderstorms and Penetrating Radiation", *Proc. Roy. Soc.* A, 130:37-63.

——— (1933): "On a Penetrating Radiation From Thunderclouds", *Proc. Roy. Soc.* A, 140:314-333.

Schove (1973): "The Biennial and Triennial Cycles of Stratospheric Reversals", *Interdisciplinary J. of Cycle Research*, 3:349-354.

Shankland, R.S., et al. (1954): "A New Analysis", *Rev. Mod. Phys.*, 27:178 —.

Shaw, H.R. (1970): "Earth Tides, Global Heat Flow, and Tectonics", Science, 168:1084; cf. (1971): ibid, *Bull. Geol. Soc. Am.*, 82:869, and (1974): "Scale Effects in the Study of Earth Tides", *Journal of Rheology.*

Simkin, T., et al. (1973): "Seismicity of a Caldera Collapse: Galapagos Islands 1968", *J. Geophys. Res.*, 78:8591.

Sleeper, H.P. (1972): "Planetary Resonances, Bi-Stable Oscillation Modes and Solar Activity Cycles", NASA Contractor Report CR-2035.

Slichter, L.B. (1972): "Earth Tides", in *The Nature of the Solid Earth*, Robertson, Ed., McGraw-Hill, NY.

Spielgelus (1627): in *Moon Madness*, by E.L. Abel, 1976.
Stilwell, G.R. & Ives, H.G. (1938): "An Experimental Study on the Rate of a Moving Atomic Clock", *J.O.S.A.* 28:226.
Stolov & Shapiro (1975): *Journal of Geophysical Res.*, May 20.
Stringfellow, M.F. (1976): "Lightning and the Sun", *Nature*, May 24.
Suguwa, C., et al. (1972): "On the Relation Between the Rotation of the Earth and Solar Activity", *Pub. Int. Latitude Obs. of Mizusawa,* Japan 8:105.
Swenson, L.S. (1972): *The Ethereal Aether*, U. of Texas Press, Austin, TX.
Tamrazyan, G.P. 7:59 ; (1967): "Tides and Earthquakes", *Icarus*, cf. *J. Geophys. Res.* 73:6013.
Taylor, L.J. & Diespecker, D.D. (1972): "Moon Phases and Suicide Attempts in Australia", *Psychological Reports*, 1972, 31:112.
Trouton, F.T. & Nobel, H.R. (1903): "The Mechanical Forces Acting on a Charged Electrical Condenser Moving Through Space", *Trans. Roy. Soc.*, 202:181; cf. Kennard, E.H. (1922): "The Trouton-Noble Experiment", *Bull. Nat. Res. Council*, 4:162-172.
Tchijevsky, A.L. (1930): "Epidemic Catastrophes and Periodic Activity of the Sun", *V.O.V.G.*, Moscow.
——————— (1964): "The Earth in the Universe", *Misl*, .Moscow.
——————— (1971): "Physical Factors of the Historic Process", *Cycles*, 22:11-21.
Tompkins, P. & Bird, C. (1973): *Secret Life of Plants*, Harper & Row, NY.
Townes, C. H., et al. (1963): "Frequency Stability of He-Ne Masers and Measurments of Length", *Phys. Rev. Let.*, 10:165-167.
Tromp, S.W. (1963): *Medical Biometeorology*, Elsevier, NY.
Vidorenkov, N.S. (1968): "The Influence of Atmospheric Circulation on the Earth's Rotational Velocity", *Soviet Astronomy AJ*, 12:303/706.
Vigier, J.P., et al. (1974): in *Astrophysical Journal Letters*: Vigier, Nottale, Karoij.
Vitels (1957): "The Solar Nature of Atmospheric Rhythms", *Trudy Central Inst Prognozov,* 51:22-43.
Vkulus & Monin (1971): "About the Spectra of Long Period Oscillations in Meteorological Fields", *Doklady Akademii Nauk SSSR*, 197,2:328-331.
Vladmirov, et ale (1970): "Solar Activity and Baric Field Oscillations in High Latitudes in the N. Hem.", *Trudy Arktich. i Antarktich. Nauchno* — Issledov. 296:160-167.
Wagner (1971): "Long Period Variations In Seasonal Sea-Level Pressure", *Monthly Weather Review,* 99,1: 49-66.
Wedler (1973): "Cycles in Atmospherics", *Interdiscip. J. Cycle Research*, 3:355-356.
Wheeler, R. (1951): "An Index of International War; an Index of Civil War", *Cycles*, 2:57-79
————— (1960): "Indexes of War", *Journal of Cycle Research*, 4:99-102.
Whitten (1975): "Preliminary Investigation of the Correlation of Polar Motion and Major Earthquakes", *Interdiscip. J. Cycle Res.*, 5:161.
Wilcox, J.M. (1968): "The Interplanetary Magnetic Field, Solar Origin and Terrestrial Effects", *Space Science Review,* 8:258.
—————— (1973): "Solar Activity and the Weather", *Stanford U. Inst. Plasma Res.*, report # 544.
Willett, H.C. (1968): "Remarks on the Seasonal Changes of Temperature and of Ozone in the Arctic and Antarctic Atmosphere", *J. Atmos. Sci.* 25:341-360.
Wolf, W., Ed. (1962): "Rhythmic Functions in the Living System", *Ann. N.Y. Acad. Sci.*, 98:753-1326.
Wood, K.D. (1972): "Sunspots and Planets", *Nature*, 240:91-93.
Zatopek, A. & Krivsky, L. (1974): "On the Correlation Between Meteorological Microseisms and Solar Activity", *Bull. Astron. Inst. Czech.*, 25, 5:257-262.
Zeeman (1915): contained in Swenson (1972).

APPENDIX II

DESCRIPTION OF CLOUDBUSTER "ICARUS"

Deleted by author. See "Note of Caution" in frontmatter of book

APPENDIX III:

BIBLIOGRAPHICAL REVIEW OF WILHELM REICH'S RESEARCH

The writings of Wilhelm Reich and his co-workers are quite voluminous, and we will use this appendix as a guide to them for the interested reader. A complete bibliography will follow this general essay. Reich began his work as a medical doctor, working as a student of Sigmund Freud. Reich soon became Freud's most promising student, leading the psychoanalytic training seminars in Vienna. Much of Reich's early writings are topically centered on psychoanalysis and medical psychiatry, with his own emphasis on the sex-economic problems of human instinctual energy. Translations of Reich's European journal articles on this phase of his work are collected in:

EARLY WRITINGS (1920-1935)

SEX-POL ESSAYS (1929-1934)

Reich, who was a prolific writer, also produced a number of books during these early years, dealing with his own unique analysis of human character structure, sexual repression, political ideology and emotional health. The following titles are English language translations of this early work:

CHARACTER ANALYSIS

INVASION OF COMPULSORY SEX-MORALITY

THE SEXUAL REVOLUTION

THE MASS PSYCHOLOGY OF FASCISM

The book PEOPLE IN TROUBLE is an autobiographical essay of Reich's work during the period 1927 - 1937 when Hitler rose to power. A collection of essays and excerpts of other pioneering works by Reich on human sexuality are contained in the volume:

GENITALITY IN THE THEORY AND THERAPY OF NEUROSIS

Reich used Freud's discoveries of infantile sexuality, the pleasure principle, and the sexual etiology of neurosis as a foundation for his own work with mental patients, which relied greatly upon these considerations. Freud was to later greatly modify and dilute his own discoveries; Reich and Freud eventually split company over irreconcilable differences in theory and therapeutic approach.

Reich's work retains in it the early discoveries of Freud, however, with important extent ions into the biophysical sphere. Three books which document Reich's work during this period of division from Freudian psychoanalysis are:

DISCOVERY OF THE ORGONE, VOL. 1: FUNCTION OF THE ORGASM

THE IMPULSIVE CHARACTER AND OTHER WRITINGS

BIOELECTRICAL INVESTIGATION OF SEXUALITY AND ANXIETY

Reich's research directions became focused more toward the questions surrounding the origin and essential nature of bodily instinctual energy, as it related to human emotion and health. A series of microbiological studies by Reich resulted in claims of the discovery of a specific "life energy" surrounding special microbial cultures. The energy was later determined to be present in all living organisms. Reich felt he had quantitatively objectified the metaphorical Freudian "libido" — he gave the energy the name orgone to preserve its identity with living organisms. These microbiological experiments, which were related to the dual problems of biogenesis and human cancer, were published in German in a book titled Die Bione. A recent English translation of this work is titled:

THE BION EXPERIMENTS

A pair of volumes which document the actual discovery of the orgone energy, starting with research on sexual dysfunction, through the microbiological and biophysical experiments, are titled:

DISCOVERY OF THE ORGONE, VOL. 1: FUNCTION OF THE ORGASM

DISCOVERY OF THE ORGONE, VOL. 2: THE CANCER BIOPATHY

The last four above mentioned titles show Reich's point of departure from traditional psychoanalysis and psychiatry toward microbiological research and biophysical experimentation, culminating in his discovery of the orgone energy.

From this point on, Reich's research was directed toward experimentation with the orgone, which was eventually found to exist also in both the atmosphere and in non-living substances. Two major books were published which explore the general relationship between the orgone energy and perception, sexuality, living organisms and the non-living world. These are:

ETHER, GOD AND DEVIL

COSMIC SUPERIMPOSITION

Reich produced a number of journal articles during this period, first in the International Journal of Sex-Economy and Orgone Research, and later in the Orgone Energy Bulletin, and Cosmic Orgone Engineering. It was in the latter of these journals that the cloudbuster device was first announced. A complete bibliography on cloudbusting is given in the body of this thesis. Books during this period of physical-atmospheric research include:

THE ORANUR EXPERIMENT, FIRST REPORT: 1947-1951

CONTACT WITH SPACE, ORANUR SECOND REPORT: 1951-1957

These were Reich's last publications. Unfortunately, due to the Food and Drug Administration's book burning activities, most of these last journals and books are still out-of-print, and not even available through interlibrary loans. Some of this material, including a bit on cloudbusting, is in the volume:

SELECTED WRITINGS

During his life, Reich experienced the bitter attack of the Nazis, the European psychoanalytic establishment, the Russian communists, the American psychiatrists, and lastly the pharmaceutical-medical-industrial complex, via the Food and Drug Administration. These attacks were never substantive, and generally took the tone of a political smear. Few, if any, of his critics seemed to have bothered to seriously duplicate the experimental work. Reich generally refused to respond to this type of pseudo-scientific criticism, choosing to ignore the attacks. He did, however, publish his own analysis of the hidden reasons for such bitter emotional reactions to his work in the books:

EMOTIONAL PLAGUE OF MANKIND, VOL. 1: PEOPLE IN TROUBLE

EMOTIONAL PLAGUE OF MANKIND, VOL. 2: THE MURDER OF CHRIST

LISTEN, LITTLE MAN!

A number of individuals have published works on Reich's life and works. Two of the best I have seen are:

WILHELM REICH AND ORGONOMY by Ola Raknes

WILHELM REICH, LIFE FORCE EXPLORER by J. Wyckoff

Reich's wife, IIse Ollendorff, has also published a biography titled:

WILHELM REICH, A PERSONAL BIOGRAPHY

A clear exposition of the Reich-F.D.A. controversy is contained in the volume:

WILHELM REICH VS THE U.S.A. by J. Greenfield

As mentioned earlier, most of Reich's journal publications, as well as those of his co-workers, were lost in the F.D.A. book-burnings. However, since 1967, a journal on Reich and Orgonomy has been published which contains up-to-date research on the orgone energy, as well as on the Reichian therapeutic techniques. Out-of-print articles by Reich and others are sometimes republished in this journal, which is:

THE JOURNAL OF ORGONOMY

For the reader unfamiliar with Reich's works, it is probably best to start by reading SELECTED WRITINGS, and the two volume set DISCOVERY OF THE ORGONE, VOL. 1: FUNCTION OF THE ORGASM, and VOL. 2: THE CANCER BIOPATHY. I believe such is a bare minimum to understand what Reich's works were about, and on what experimental basis his claims were laid. The reader could then fill in, gaps with more specific information from the other works of Reich and his co-workers mentioned in this synopsis.

Republished Works by Wilhelm Reich, M.D.
Noonday Press paperback: Farrar, Straus and Giroux hardback:

THE BION EXPERIMENTS
THE CANCER BIOPATHY (Discovery of the Orgone, Vol.2)
CHARACTER ANALYSIS
THE EARLY WRITINGS OF WILHELM REICH: VOL.l, 1920-35
ETHER, GOD AND DEVIL / COSMIC SUPERIMPOSITION
FUNCTION OF THE ORGASM (Discovery of the Orgone, Vol.l)
GENITALITY IN THE THEORY AND THERAPY OF NEUROSIS
THE INVASION OF COMPULSORY SEX-MORALITY
LISTEN, LITTLE MAN!
THE MASS PSYCHOLOGY OF FASCISM
THE MURDER OF CHRIST (Emotional Plague of Mankind, Vol.2)
PEOPLE IN TROUBLE (Emotional Plague of Mankind, Vol.l)
REICH SPEAKS OF FREUD
SELECTED WRITINGS
THE SEXUAL REVOLUTION
THE IMPULSIVE CHARACTER AND OTHER WRITINGS
SEX-POL ESSAYS, 1929 – 1934

Books About Wilhelm Reich (various authors)

Baker, E.F. (1967): MAN IN THE TRAP, Macmillan, N.Y.
Eden, J. (1972): ORGONE ENERGY, Exposition Press, N.Y.
Greenfield, J. (1974): WILHELM REICH VS. THE U.S.A., W.W. Norton, N.Y.
Ollendorff, I. (1969): WILHELM REICH, A PERSONAL BIOGRAPHY, St. Martin's Press, N.Y.
Raknes, O. (1970): WILHELM REICH AND ORGONOMY, St. Martin's Press, N.Y.
Reich, P. (1973): A BOOK OF DREAMS, Harper and Row, N.Y.
Ritter, P., Ed. (1958): WILHELM REICH MEMORIAL VOLUME, Ritter Press, Nottingham, England
Wyckoff, J. (1973): WILHELM REICH, LIFE FORCE EXPLORER, Fawcett Premier, Greenwich, Conn.

Unavailable, Out Of Print Works by Wilhelm Reich, Et Al
* asterisk indicates a periodical.

*INTERNATIONAL JOURNAL OF SEX-ECONOMY AND ORGONE RESEARCH, Volumes 1 - 4, 1942 - 1945, Theodore P. Wolfe, M.D., Editor, Orgone Institute Press, N.Y. Official organ of the International Institute for Sex-Economy and Orgone Research, Wilhelm Reich, Director.

*ANNALS OF THE ORGONE INSTITUTE, Volume 1, 1947, Other volumes reprinted in Ether, God and Devil

*ORGONE ENERGY BULLETIN, Volumes 1 - 5, 1949 - 1953, Edited by the teaching and research staff under the direction of Wilhelm Reich, Orgone Institute Press, N.Y.

THE ORGONE ENERGY ACCUMULATOR, ITS SCIENTIFIC AND MEDICAL USE, 1951, Wilhelm Reich Foundation, Orgone Institute Press, Orgonon, Rangeley, Maine

HISTORY OF THE DISCOVERY OF THE LIFE ENERGY (Many volumes never released, presently in storage at Reich archives, Rangeley, Maine)
Doc. Vol. A-IX-B, BIBLIOGRAPHY ON ORGONOMY, 1953
Doc. Vol. A-XI-E, THE EINSTEIN AFFAIR, 1953
Doc. Sup. A-XII-EP, RESPONSE TO IGNORANCE, 1955
Doc. Vol.A-XII-EP , CONSPIRACY, AN EMOTIONAL CHAIN REACTION, 1954

*COSMIC ORGONE ENGINEERING (CORE), Volumes 6 and 7, 1954 - 1955; continuation of Orgone Energy Bulletin

THE ORANUR EXPERIMENT, FIRST REPORT: 1947-1951, 1951, Initially published as Orgone Energy Bulletin, Vol. 3, #4, partially reprinted in Selected Writings

CONTACT WITH SPACE, ORANUR SECOND REPORT: 1951-1956, 1957, Core Pilot Press, N.Y.

RECORD APPENDIX, VOL. III: SUPPRESSED DOCUMENTARY EVIDENCE, 1957

*ORGONOMIC MEDICINE, Volumes 1 and 2, 1955-1956 Forerunner of the Journal of Orgonomy

APPENDIX IV

DATES OF CLOUDBUSTER TESTS AND DAILY WEATHER MAPS

The following are dates during which tests were made for:

1. Cloud Dissipation (20 total)

1977:	18 June	1978:	1 April
	16 July		8 April
	14 August		23 April
	16 August		14 May
	20 August		27 May
	1 October		20 June
	8 October		30 June
	31 October		24 July
	16 December		25 August
			8 September

Precipitation Abatement (3 total)

1977:	25 June	1978:	16 April
			23 May

Precipitation Enhancement (12 total)

1977:	14 July	1978:	26 August
	17 July		3 September
	30 July		10 September
	8 September		17 September
			24 September
			1 October
			8 October
			15 October

This appendix originally contained copies of Daily Weather Maps for North America, reflecting conditions on the morning of and the morning after each individual precipitation abatement and enhancement test (2 maps per test). They were for 7:00 am Eastern Standard Time (6:00 am Central Standard Time) for the day in question. Unfortunately, the quality of those materials has degraded, such that reproduction in the present publication provided nothing of useful value. Consequently, they are not included.

APPENDIX V

SELECTED LOG NOTES FOR PRECIPITATION ABATEMENT AND ENHANCEMENT OPERATIONS

Following now are excerpts from log notes for a number of selected cloudbuster tests, one date for precipitation abatement, and three dates for precipitation enhancement. These were typed up directly from my original written field operations log-books.

It may be noted that, during a few of these operations, the storm systems which followed also initiated tornado watches, severe storm watches, and flash flood watches on the part of the National Severe Storms Forecasting Center. In all cases, however, severe weather failed to significantly materialize. This suggests, the weather forecasters were caught "off-guard" by storms which developed contrary to their expectations, and it consequently prompted their rational, but overly-cautious public warnings. As detailed in Chapter V., on "Cloudbuster Operations and Severe Storms", the Kansas operations were followed by a net reduction in severe weather.

LOG NOTES: PRECIPITATION ABATEMENT TEST

OF 16 APRIL 1978

<u>16 April 1978</u>
Afternoon draw to STOP rain: flood warnings out.
Many low pressure centers to the west
OROP[1] to reverse E. to W. OR flow back w. to E.

3:45 pm -grey, overcast, wind from East, drizzle
"Flash flood watch tonight" WDAF, K.C.

<u>Operational Plans:</u> attempt will be made to stop rain by drawing for 15 -20 minutes *from* West, specifically:

1. OR concentrations to West of site

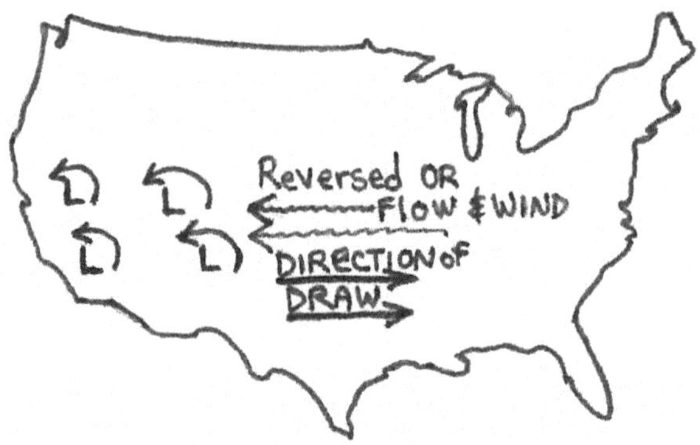

2. Wind flow for last two days from east to these areas building up potential

3. Draw against E. to W. flow to reverse buildup to W. and restore W. to E. flow this should break up low pres. centers to West of here; these could come through in succession and devastate the area.

<u>Possible Danger:</u> buildup of potential to East of site

---

1. OROP is a slang designation meaning "orgone operation"; similarly OR means "orgone".

4:00 pm -start OROP

4:05 pm -stop OROP — relay failure!!!

4:15 pm -restart OROP, use manual controls

4:20 pm -bird activity, clouds breaking up a bit
-draw to S.W. seems to abate wind
-sky now opening up to S.W. more than elsewhere, but it will not yield
-no more drizzle

4:55 pm -stop OROP, sky still overcast but wind abated 50%, wind still from East

5:10 pm -no change in sky, still socked in

5:50 pm -no change in sky, still socked in

6:10 pm -sky opening up to S.W. and South

6:15 pm -sky opening up overhead OROP site — blue sky visible in holes overhead, which open and shut, but S.W. remains clear

Night of 16th — I hear distant thunder, slight drizzle cloudy, — I can't sleep, arm is hurting bad from manual operation — I got a bad charge. Voltmeter shows intense peak around 9:30 pm.

Morning of 17th — Thunder still apparent at distance slight drizzle stops, birds awake and chirping all night long.

6:30 am -sky alternates from clear blue to cloudy - like it can't make up its mind - surface wind variable

8:30 am -sky blue with high cirrus cool 52'
-Nat. Weather Servo says 100% chance of rain + flash flood watch.
-"A day which does not look like 100% chance for rain" KANU radio
-"Big thunderstorm came out of west late afternoon (yesterday), really looked bad! Drizzled for awhile and then nothing - Really confused last night!" — local meteorologist

8:45 am -sky does not know what to do — contract down and then expands out again? Or is the pulse a favorable sign?

9:10 am -National Weather Service now forecast 30% chance of rain, 80% for later today
-Light rain in Topeka (to West) heading E.N.E.
-8' increase in temperature in last 2 hours

9:30 am -"A persistent low in Texas panhandle (to S.W) filled in during early morning" N.W.S.
-sky now overcast

2:3 0 pm -sky socked in, hazy

6:30 pm -Drizzle here so far but no heavy cloudbursts; around 5:00 pm the TV weather showed a line of storms oriented N-S just west of Kansas City heading East — flash flood and tornado watches are out

9:30 pm -sky crystal clear in Lawrence

10:30 pm -Kansas City: "not much rain" 1/100", even
relative humidity is 95%
-tornado watch cancelled
-flash flood watch cancelled
-funnel clouds spotted in Nebraska and Oklahoma but none in Kansas — no damage
It seems that the potentially flooding weather system passed over the area as if "holding its breath" and did not dump its load of precipitation.

18 April

afternoon: day cloudy but crisp, clear and cool, wind from west
no flood in Kansas City or N.E. Kansas

Retrospect: Kansas City missed a flood but West Kansas got dust storms. Would these dust storms have been avoided if rains had continued their march? See Figure 28 in body of thesis

LOG NOTES: PRECIPITATION ENHANCEMENT TEST OF 26 AUGUST 1978

<u>26 August 1978</u>
OROP to break drought tendency which has extended S.E. to Wichita and S.W. Kansas — no rain there, little rain here. Many crickets, grasshoppers have been plaguing the area of the USA for some time; locusts are on the move in Africa also.

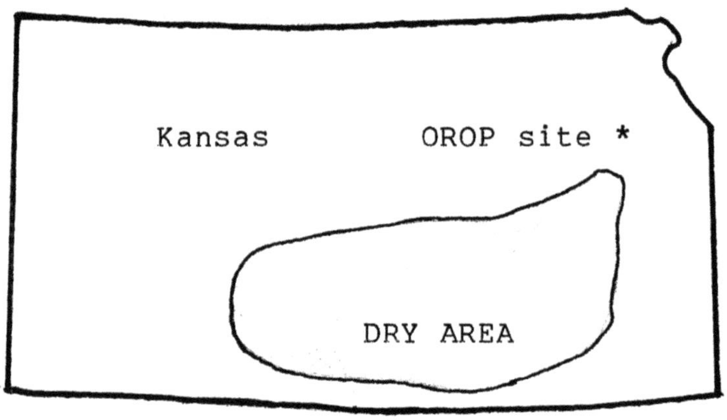

<u>Operational Plans:</u>

1. Draw off and neutralize DOR to S.W. and W. of OROP site.

2. Contract OR envelope with zenith draw

3. Bring in fresh OR and fronts from N.W. and N. via compass draw to N.W.; fronts there have been "stagnated" for days (Nat. Weather Serv.)

12:15 pm -no bird noises or birds seen in sky
-wind about 10 mph from South
-sky hazy, a few high cirrus wisps, expanded jet contrails; a single cumulus building on the N.E. horizon

12:15 pm -commence OROP: draw to S.W.

12:25 pm -commence zenith draw

12:30 pm -cloud character noticeable changes
-sky filling in between cirrus, which has spread outward
-wind is steady now at ground level, was gusty before

12:40 pm -Cumulus forms springing up to S. of cloudbuster

NOTE: Sky filling in with altostratus or cirrus in between areas of jet contrails. Sky continuing to fill in at high levels.

NOTE: Since I am operating during the time of daily maximum atmospheric excitation (solar noon), I do not expect to see results from this operation until late this evening, when the earth OR envelope naturally contracts down all on its own. I must be careful not to overdraw as results will come late anyway.

1:15 pm -cumulus cover increasing

1:45 pm -cumulus clouds all around now; altostratus still present, but holes in it up to blue sky

2:00 pm -stop zenith draw
-draw now to N. and N.W. to break loose fronts in that area and to counter flow from S. Hopefully this will reinforce OR contractive processes.

2:05 pm -cumulus in area of cloudbuster have dissipated but not those toward the horizon. Altostratus now intensifying.

2:11 pm -Zenith draw again
-now cumulus grow as altostratus declines!!

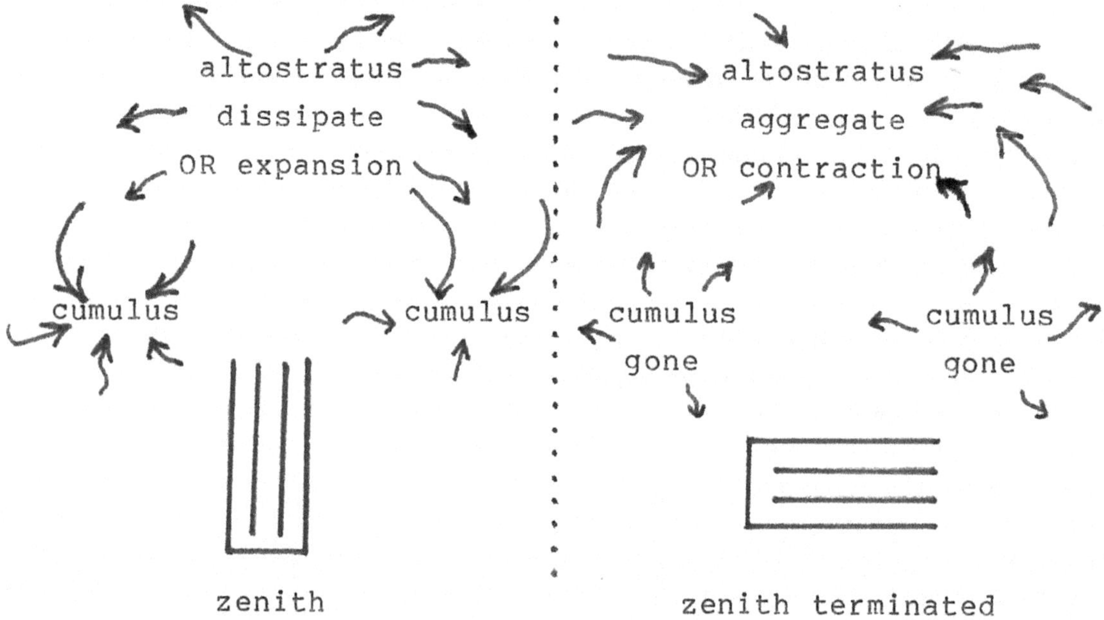

"spring" in atmosphere at work — expands, the contracts

2:20 pm -terminate zenith draw
-start N.W. draw again

2:30 pm -terminate OROP:
-sky now 85% covered with altostratus
-large building cumulus scattered over area
-many birds in sky now, active

NOTE: Decided to leave two tubes of cloudbuster pointed toward the N.W. with partial grounding ("slow draw") I will shut them down also depending upon the 5:30 pm weather report

6:15 pm -TV weather report:
"Severe storm watch out for N.E. Kansas; 60% rain chance; winds from NORTH!! Storms in Topeka! A new low pressure area seen on weather map"

6:50 pm -close up cloudbuster
-raining all around cloudbuster site
-wind from N.W.
-large collar cloud seen in area, but it dissipated as it approached the cloudbuster site!

<u>27 August 1978:</u> morning
- wind from N. W.
-clean and cooler, crisp .
-28 and 29 August retain this weather pattern

Spoke to B.S. in Wichita: "Storms came out of blue!", "not anticipated", "two previous months dry and hot", "All farm ponds full in area as seen from aircraft"

NOTE: Tropical Depression (Debra) was initiated in Gulf of Mexico at same time OROP began??!!

NOTE: On zenith draws there seems to be a distinct pressure buildup - subjective feeling - while OROP is in progress. This is relieved after zenith draw is terminated and is in agreement with the diagram on the previous page. Will this register on a barometer?

Retrospect: A significant increase in precipitation occurred over the state, beginning just after commencement of this operation. Significant precipitation occurred in the area of the small sketch map here labeled as "dry", in accordance with operational plans. See Figure 36 and satellite Photo Sequence 11 in body of thesis.

LOG NOTES: PRECIPITATION ENHANCEMENT TEST OF 3 SEPTEMBER 1978

<u>3 September 1978</u>

Before OROP: winds 5 - 10 MPH from East
no clouds in sky

(NOTE: During this operation I was able to confirm that barometric pressure was altered, as per the subjective sensations noted in the operation of 26 August 1978. This phenomena is dealt with in the body of the thesis, in Figure 30, and will not be repeated here substantially, even though it appears in the original log notes.)

Weekly OROP: using primarily zenith and N.W. Draws

12:55 pm -OROP start
-scan draw from W. to N.W. to N. back and forth, to mobilize OR energy

1:00 pm -start zenith draw

1:05 -bands of haze setting in overhead

1:25 -subjective sensation of pressure buildup again, in head
-wind shift to S.E.

1:40 pm -draw to S.W. (to mobilize galactic OR stream)
-haze seems to intensify while doing this

1:45 pm -back to zenith draw again — PATIENCE!! It is so clear and blue today I question whether or not it can be made to rain today

1:46 pm -haze now coalescing into small cumulus!!
-more cumulus now visible to S.W. and W. of OROP site! Growing in size

2:08 pm -first bird seen in sky, soaring high
-cumulus continue to form and dissipate, but no really large ones seen yet

2:40 pm -terminate zenith draw, start S.W. draw again
-many soaring birds now seen to S.W.

2:55 pm -Stop OROP — watch ...
-cumulus building to east of site
-barometric pressure is dropping rapidly!!
-pressure sensation in head is letting up also!

3:05 pm -cumulus continue to build mostly to N. and E.

3:30 -cumulus buildup seems to have stopped

3:35 pm -Restart OROP — zenith draw

3:45 pm -thunderhead tops now visible to S.W.!
-towering rapidly

3:50 pm -Stop zenith draw
-Start slow draw to N.W. with 1/2 grounding and 2 tubes open

4:00 pm -wind from S.E.
-more thunderheads on S.W. horizon, slowly building and approaching

4:55 pm -thunderheads seem to be moving out of area
-start zenith draw again

5:20 pm -Stop zenith: draw to s.w. with 2 tubes (try
to get storm to south to move N.E.)

6:00 pm -sky cloud-free except to S.W. which has cirrus
and thunderhead tops spread across it

6:20 pm -terminate OROP
-large bank of clouds to S.W.: may build up and
move into this area or may dissipate or
persist there. We will see.
-A drive to South of Lawrence shows these to
be well-developed "supercell" thunderstorms

9:00 pm -thunderstorms still visible way off to south
of Lawrence; some lightning visible

10:00 pm TV Weather:
"heavy showers with some hail to S.W. around
Emporia area. Contradictory reports on wind direction:
E., N. and S." Atmosphere trying to make up mind??

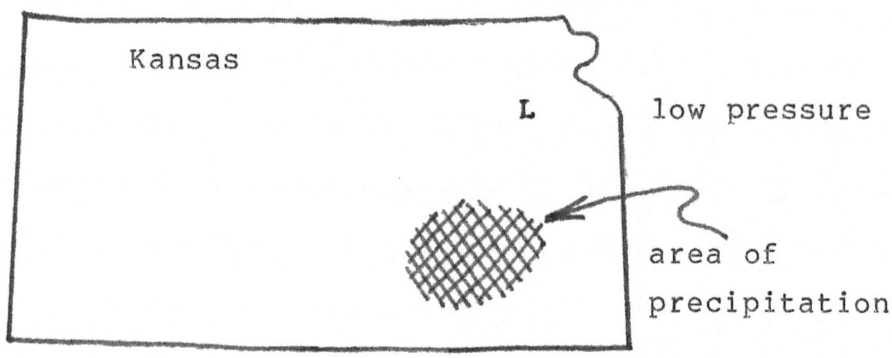

<u>4 September 1978:</u> morning is cool and crisp, but hot
and muggy by afternoon

Evening TV weather report:

Radar image of big storms and rain centers to S. and S.E. These are the same storm centers that appeared yesterday during the OROP.

6 :30 pm - I drove out to the operational site with the intention of attempting to move this rainy area North. I got there and looked at the S.E. horizon, to see immense cloud tops — supercell thunderheads in a big row. I got the feeling these storms were building up until they could overcome the OR expansive tendency existing to the north of them, at the OROP site. They could not move north until they built up enough. I have decided to do nothing but sit and watch for the rest of the evening.

Retrospect: These storms pushed northward during the evening, bringing light rains over E. Kansas. The cloud bottoms were at an abnormally high altitude, giving the impression that they had to "ride up and over" an area of still- existing OR expansion. Precipitation was light, but these storms produced much lightning. See Figure 37 and satellite Photo Sequence 12 in the body of the thesis.

LOG NOTES: PRECIPITATION ENHANCEMENT TEST OF 17 SEPTEMBER 1978

17 September 1978
Weekly OROP: zenith + N.W. draw

1. start with N.W. draw
2. then zenith draw
3. N.W. draw again

12:15 pm -sky overhead 60% overcast, stratiform clouds
-wind to E. or S.E.

12:20 pm -Start OROP, N.W. draw
-clouds begin to change in character from stratiform to cumulus with elevating tops

12:25 pm -atmosphere has been mobilized
-clouds moving in from South
-bunching up

12:35 pm -change to zenith draw

NOTE: During zenith draw, when overhead clouds begin to break up and dissipate, it does not necessarily mean that a pattern of widespread dissipation has set in. In this case we see larger cumulus converting to thunderheads on the horizon. "Spring" of atmosphere re-contracts OR envelope once OROP is terminated.

12:45 pm -stop zenith draw
-start N.W. draw

12:55 pm -stop OROP — observe

1:05 pm -sky has now changed back to like it was before OROP began — a temporary state?

1. Before OROP 2. During OROP 3. After OROP

NOTE: I left my watch at home today and notice that I am now watching the sky more than before: at earlier OROPs I watched the clock and "timed" operations rather than gauging the operation by what OR in the sky was doing. I must leave my watch "at home" more often!

1:20 pm -zenith draw again

1:30 pm -zenith draw seems to fill in holes overhead
but there is no observed thunderhead
activity anywhere

1:35 pm -Pattern to E. of OROP site now changing to
thunderheads!
-cumulus forming to S. and S.W.; hard to tell
if they are small and close or large and
far off

1:45 pm -overhead cloud movement has stopped greatly

1:50 pm -change to N.W. draw
-almost immediately back to stratus again!

1:55 pm -zenith again!

2:10 pm -sky clearing overhead
-cumulus again sprouting here and there

2:12 pm -draw now to N.E., against galactic stream
-thunderheads now sprouting to S.E.

2:25 pm -zenith again: large thunderheads to E. and S.E., they seem to "blossom" exactly after zenith draw commences!
-cumulus overhead now organizing into streets
-birds high in sky overhead

2:45 pm -Stop OROP

3:00 pm -much vertical mixing, cumulus streets, birds
-time will tell if this operation proceeds toward rain

4:30 pm -sky in Lawrence is overcast heavily with altostratus and a few cumulus
-National Weather Service radio: "Severe storm watch and tornado watch until 5:00 pm!! Flash flood watch tonight for Kansas City and St. Joseph!!"

5:30 pm -TV weather map:
-Tornado watch and flash flood watch for Kansas City until 11:00 pm tonight!!!

(see map on next page)

6:30 pm -large black clouds overhead Lawrence

10:00 pm -local heavy showers, intermittent
-tornado watch cancelled (funnel cloud spotted earlier but did not touch down. Yesterday, the day before this operation, a killer tornado swept through Grinell, Iowa)
-stationary front expected to persist through the week
-new line of storms generated in central KS
-4" rain in Kansas City

5:30 pm TV weather Map:

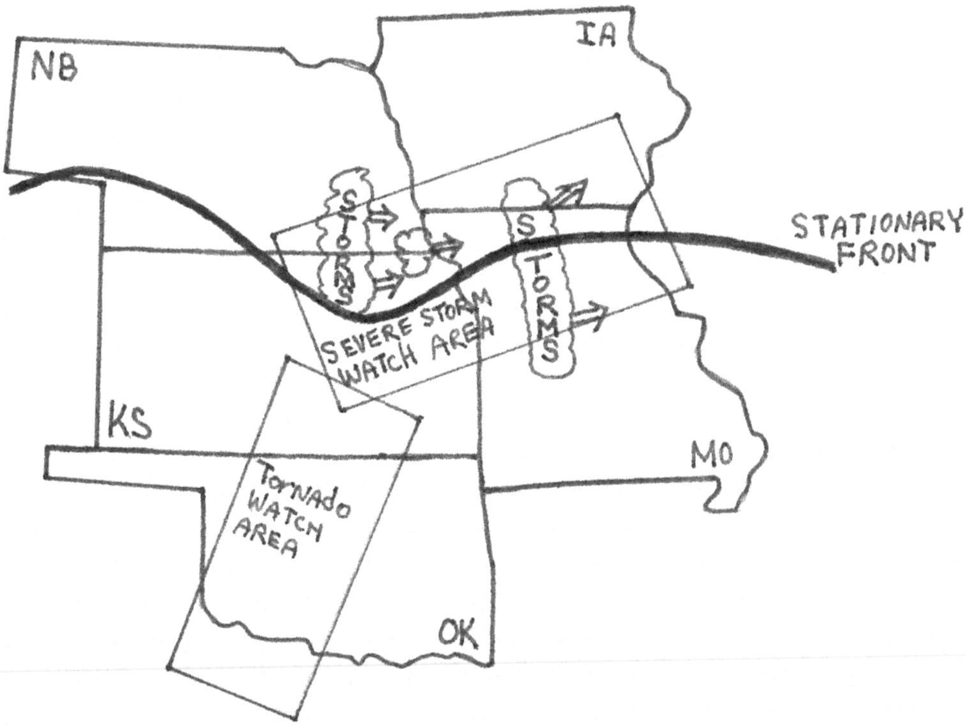

12:00 am -midnight:

-no more storms over Kansas City
-storms over Missouri diminishing
-NO MAJOR DAMAGE ANYWHERE, only tree limbs, garbage cans, etc.

<u>Retrospect:</u> This operation generated rainfall over the entire state of Kansas in significant quantities. No damage, other than broken tree limbs, plus moved debris and litter -- garbage cans, etc. -- were reported. See Figure 39 and Satellite Photo Sequence 14 in the body of the thesis.

About the Author

James DeMeo, Ph.D., is Director of the Orgone Biophysical Research Lab, which he founded in 1978. He received his doctorate in Geography from the University of Kansas, where his research corroborated various aspects of Wilhelm Reich's social and biophysical discoveries. He also studied chemistry and environmental science at Florida International and Florida Atlantic University. Dr. DeMeo was formerly on the faculty of Geography at Illinois State University and the University of Miami. His interdisciplinary research ranges across social/cultural and biophysical subjects, including cross-cultural, historical studies on the effects of drought and desertification on the origins of warfare and social violence, with laboratory experiments on cosmic cycles and life-energy subjects, and field research and applications of the Wilhelm Reich cloudbuster for drought-abatement and greening of drylands. He has published over 100 journal articles and book-chapters on the issues of energy resources, health, cultural history, environmental problems, and experimental orgone biophysics. He is author of the popular *Orgone Accumulator Handbook* and of *Saharasia, The 4000 BCE Origins of Child-Abuse, Sex-Repression, Warfare and Social Violence, In the Deserts of the Old World*. He is also Editor of *Heretic's Notebook: Emotions, Protocells, Ether-Drift and Cosmic Life Energy*, and *On Wilhelm Reich and Orgonomy*, and Co-Editor of the German-language work *Nach Reich: Neue Forschungen Zur Orgonomie*. He lives in the high-altitude mountains of the Southern Oregon Siskiyou Range, where he maintains a private laboratory and educational facility.

www.ingramcontent.com/pod-product-compliance
Ingram Content Group UK Ltd.
Pitfield, Milton Keynes, MK11 3LW, UK
UKHW051029041026
15470UKWH00015B/1749

9 780962 185540